W9-AZY-046

A GUIDE TO BIOMOLECULAR SIMULATIONS

FOCUS ON STRUCTURAL BIOLOGY

Volume 4

Series Editor
ROB KAPTEIN
Bijvoet Center for Biomolecular Research,
Utrecht University, The Netherlands

A Guide to Biomolecular Simulations

by

OREN M. BECKER
Predix Pharmaceuticals, Ramat-Gan, Israel

and

MARTIN KARPLUS
Harvard University, Cambridge, MA, U.S.A. and
Université Louis Pasteur, Strasbourg, France

A C.I.P. Catalogue record for this book is available from the Library of Congress.

ISBN-10 1-4020-3586-1 (HB)
ISBN-10 1-4020-3587-X (e-book)
ISBN-13 978-1-4020-3586-9 (HB)
ISBN-13 978-1-4020-3587-6 (e-book)

Published by Springer,
P.O. Box 17, 3300 AA Dordrecht, The Netherlands.

www.springer.com

Printed on acid-free paper

All Rights Reserved
© 2006 Springer
No part of this work may be reproduced, stored in a retrieval system, or transmitted
in any form or by any means, electronic, mechanical, photocopying, microfilming, recording
or otherwise, without written permission from the Publisher, with the exception
of any material supplied specifically for the purpose of being entered
and executed on a computer system, for exclusive use by the purchaser of the work.

Contents

Preface   vii

Introduction: Note to the Student   1

Introduction: Note to the Instructor   3

Introduction: UNIX   4

Introduction: CHARMM Primer   10

Introduction: CHARMM Template Files   18

Lab 1:   Introduction to Molecular Visualization   19

Lab 2:   Energy and Minimization   35

Lab 3:   Minimization and Analysis   51

Lab 4:   Conformational Analysis   65

Lab 5:   Basic Molecular Dynamics in Vacuum and in Solution   75

Lab 6:   Molecular Dynamics and Analysis   101

Lab 7:   Ligand Dynamics in Myoglobin   131

Lab 8:   Normal Mode Analysis   155

Lab 9:   Free Energy Calculations   173

Lab 10:   Minimum Energy Paths   193

Lab 11:   Multiple Copy Simultaneous Search   205

Lab 12:   Hemoglobin Cooperativity: the T-R Transition   213

References   221

Index   229

Preface

The first molecular dynamics simulation of a macromolecule of biological interest was published in 1977. The simulation concerned the bovine pancreatic trypsin inhibitor (BPTI), which has served as the "hydrogen molecule" of protein dynamics because of its small size, high stability and relatively accurate x-ray structure, available in 1975; interestingly, its physiological functions remain unknown. Although this simulation was crude by present-day standards, the results were instrumental in replacing our view of proteins as relatively rigid structures with the realization that they were dynamic systems, whose internal motions play a functional role. Over the intervening years, such simulations have become a central part of biophysics.

Simulations can provide the ultimate detail concerning individual particle motions as a function of time, so they can be used to answer specific questions about the properties of a model system, often more easily than experiments on the actual system. A significant aspect of simulations is that, although the potentials employed in simulations are approximate, they are completely under the user's control, so that by removing or altering specific contributions their role in determining a given property can be examined. This is most graphically demonstrated by the use of 'computer alchemy' – transmuting the potential from that representing one system to another during a simulation – in the calculation of free energy differences.

The growth in the number of studies using molecular dynamics for biomolecular simulations has been fueled by the general availability of programs and the computing power required for meaningful studies. The first simulation was less than 10 ps in lengths, but it is now de rigueur to do simulations that are at least 1000 times as long (10 ns); they, nevertheless, take a factor of about 50 less time for the same sized system, though of course much larger systems (10^4 to 10^6 atoms instead of 500) are sometimes necessary to study if an explicit solvent and/or membrane environment is included. Another important consequence of the access to faster computers is the fact that multiple simulations can be done to obtain converged results with statistical error estimates.

The number of publications using molecular dynamics is now in the thousands (A search with "molecular dynamics" and "proteins" as key words of the ISI "Web of Science" [http://www.isinet.com/isi/products/

citation/wos/] yielded 2337 papers published since the beginning of the new millenium (January 2000).),

Applications of molecular dynamics in biophysics range over many areas. They are used in the structure determination of macromolecules with x-ray and NMR data, the modelling of unknown structures from their sequence, the study of enzyme mechanisms, the estimation of ligand-binding free energies, the evaluation of the role of conformational change in protein function, and drug design for targets of known structures.

The widespread application of molecular dynamics and related methodologies suggests that it would be useful to have available an introductory self-contained course by which students with a relatively limited background in chemistry, biology and computer literacy, can learn the fundamentals of the field.   This "*A Guide to Biomolecular Simulations*" tries to fill this need.   As the basic learning tool, the exercises in the Guide are based on the CHARMM program, which is one of the most widely used programs in the field.   CHARMM was selected because it is the program with which we do our research.   To graphically illustrate the results we have used a version of QUANTA, which is distributed by Accelrys, Inc.

The *Guide* consists of six chapters which provide the fundamentals of the field and six chapters which introduce the reader to more specialized but important applications of the methodology. For the latter, a selection was made, based primarily on our experience.

We hope that students who work through this *Guide* will be able to go on to read papers in the field with a good level of understanding and to initiate their own studies of fundamental and applied problems in the rapidly growing area of molecular dynamics simulations. References are given, many to the original work in a given area. Using them as a key, the student will be able to search for more recent work on the Web.

Many people have contributed to making this *Guide* a reality. We thank them in the Acknowledgements.

ACKNOWLEDGEMENTS

The material on which the *Guide* is based was developed at Harvard University during the 1990's over several years as a companion "laboratory" for the biophysics course "Molecular Biophysics Simulations of Macromolecules" (Biophysics 164), the lectures of which were given mainly by one of the authors. The original course did not have a laboratory and was taught jointly by Steve Harrison and Martin Karplus. After the latter became responsible for the course in its entirety, a "laboratory" was introduced in which students were able to do computer calculations that illustrated the points made in the lectures.

The first version of the laboratory manual was prepared by Herman van Vlijmen, based on the exercises developed by Postdoctoral Fellows from the Karplus group. Many of the original contributions were extended and modified over the years. The list of the original contributors:

Lab 1: Herman van Vlijmen, Oren M. Becker, Yael Marantz
Lab 2: Diane Joseph-McCarthy, Carla Mattos
Lab 3: Annick Dejaegere, Roland Stote
Lab 4: Annick Dejaegere, Roland Stote
Lab 5: Oren M. Becker
Lab 6: Jeffrey D. Evanseck, Masakatsu Watanabe, Oren M. Becker
Lab 7: Leo S. D. Caves
Lab 8: Oren M. Becker
Lab 9: Stefan Boresch, Oren M. Becker
Lab 10: Oren M. Becker
Lab 11: Amedeo Caflisch
Lab 12: Leo S. D. Caves

We would also like to thank Ryan Bitetti-Putzer, Aaron Dinner, Carla Mattos, Sung-Sau So, Herman van Vlijmen and Ori Kalid for reading through the *Guide*, testing it, and offering valuable comments.

The goal of the computer labs you are about to perform is to make you proficient with computational tools available for studying proteins and other biomolecules. Specifically, we will focus on molecular visualization programs (e.g., InsightII and QUANTA) and on the molecular mechanics program CHARMM. During the labs you will learn how to use these programs and will become acquainted with many computational procedures and techniques.

While you are not expected to become an expert in any of these programs or technique, you are expected to understand what they do and to be able to carry out basic tasks by yourself. To help you reach this goal many of the labs include *template command files*, which require only small modifications before the respected calculations can be performed. Modifying these files will give you first hand experience with the details of the computational techniques, without requiring high-level expertise.

Here are a few *guidelines* that will help you benefit from the lab exercises:

* It will be to your advantage to read through the lab instructions *before* tackling the exercise!
* In labs where you have to modify CHARMM *template files*, read through each file very carefully and make sure you understand what every command-line performs.
* There are questions scattered throughout the lab instructions. Answer the questions as you work your way through the lab instructions.

To make the labs clearer the following general conventions are used:

bold-face	UNIX, InsightII and QUANTA commands
italics	directory-names and file-names
CAPItals	CHARMM commands

The material is divided into two parts: Basic techniques and Advanced topics. It is suggested that students performs all the basic labs, and then select several advanced topics according to their inclination.

1

BASIC LABS

Lab 1 -   Introduction to Molecular Visualization
Lab 2 -   Energy and Minimization
Lab 3 -   Minimization and Analysis
Lab 4 -   Conformation Analysis
Lab 5 -   Basic Molecular Dynamics in Vacuum and in Solution
Lab 6 -   Molecular Dynamics and Analysis

ADVANCED LABS

Physical/Computational focus:
Lab 7 -   Ligand Dynamics in Myoglobin
Lab 8 -   Normal Mode Analysis
Lab 9 -   Free Energy Calculations
Lab 10 -   Minimum Energy Paths

Biophysical focus:
Lab 11 -   Multiple Copy Simultaneous Search
Lab 12 -   Hemoglobin Cooperativity: the T-R transition

This *Guide* was tested for CHARMM 27, InsightII 2000 and QUANTA 2000 running on UNIX workstations.

The labs are fully compatible with earlier version of CHARMM. They should also be compatible with the earlier version of InsightII and QUANTA, although in this case some variations are to be expected.

<u>Before starting</u> the course you will have to take care of the following:

1. Copy the *Guide* directory structure from the distribution CD to a common accessible location.

2. Set a shell environment variable LAB to hold the path to the *Guide* directory on your computer (**setenv** LAB <lab path>). The <lab path> will now be referred to in this *Guide* by $LAB.

3. Set up login accounts for all students (or groups of students) on the appropriate UNIX workstation(s).

4. Make sure that each student account has the following definitions in the *.login* file (alternatively in the *.cshrc* or *.profile* files):

   source */usr/msi/quanta2000/.setquanta*   [if using QUANTA]
   source */usr/msi/cshrc*   [if using InsightII]
   setenv LAB <lab path>
   set charmm <charmm executable>

UNIX is the operating system used by most workstation computers, including the ones you will be working with. This means that you have to use UNIX commands to do things. To help you, the most essential UNIX commands will be briefly described in this section. This will enable you to get started. During the course you will probably learn more about other useful commands.

GETTING STARTED

passwd

Login onto the computer with your user name and password (if you do not have them yet a Teaching Assistant will give them to you). Once you are logged in you may change your password, if desired, by typing **passwd**, and replying to the prompt for a new password.

mkdir

To keep your files in order it is best to create a sub-directory for every lab. Use the **mkdir** command to create a new sub-directory. For example, if you want to create a new sub-directory named *lab1* type: **mkdir** *lab1*.

cd

After creating a sub-directory the next step is to move to that sub-directory. The command for changing directories is **cd**. For example, to move to the lab1 sub-directory type: **cd** *lab1*. You can always go back to your home directory just by typing **cd**.

PATH

Files are specified by a file-name and by a "path", which defines the file's location in the hierarchical structure of the directories. The path can be "full"; i.e., specified from the *root* directory (indicated by a / at the beginning of the path), or it can be relative to the current directory. For example:

pdb8lyz.pdb	A file in the current directory.
/usr/bp151/lab1/pdb8lyz.pdb	A file specified with its full path.
lab3/LYZ_MIN1.CRD	A file specified with a relative path.

BASIC COMMANDS

Some useful symbols that are used while working with UNIX are:

4

SYMBOLS	.	The current directory.
	..	The directory directly above the current one.
	*	A wildcard, stands for any symbol or sequence of symbols.

BASIC
UNIX

A few commands that are useful in manipulating files are:

cat	Display the contents of a text file.
cp	Copy a file to a different directory or to a different filename.
head	Show the first few lines of a file (the number of lines can be specified following a - sign, see example below).
ls	List the contents of a directory.
mkdir	Create a new directory.
more	Display the contents of a text file page by page. To page through the file press the space bar; to quit press "q".
mv	Change a filename or move the file to a different directory.
pwd	The name of the current directory.
rm	Remove a file.
tail	Show the last few lines of a file (the number of lines can be specified following a – sign, see **head** example below).

Examples:

cp * *lab1/.*	Copy all the files in the current directory to the *lab1* sub-directory.
mv *.pdb* ../.	Move all the files that end with the suffix ".pdb" from the current directory to the directory above it.
rm *pep**	Remove from the current directory all files that start with the sequence "pep".
cd ..	Change to the directory directly above the current directory.
head -10 *4pti.pdb*	Show the first 10 lines of the file *4pti.pdb*.

DIRECTING DATA & COMMAND PIPES

When running programs it is often useful to direct the *input* or the *output* of the program. This action is done by the > and < operators. For example, if you want the input to a certain **charmm** program to be read from an *input_file* and the output to be stored in an *output_file*, type:

charmm < *input_file* > *output_file*

PIPE

Another powerful UNIX capability is the possibility of pipelining commands. This means that you can directly use the output of one command as an input for the next command, without having to save it to a file. This is done with the pipe sign |.

Examples:

ls .. | **tail** -2

list the content of the parent directory (..) but display only the last two lines.

charmm < *input_file* | **head –20**

run a **charmm** program using input from the file *input_file* and display <u>only</u> the first 20 lines of output on the screen.

cat *my_file* | **more**

identical to the **more** command.

CONTROLLING JOBS

When you execute a program (e.g., run a simulation) you are creating a "job" on the computer. There are many UNIX tools to control this job; the most basic ones are listed below.

^z

bg
fg

The first thing that is important to realize is that a job can run either in the *foreground* or in the *background*. When a job runs in the foreground it fully occupies the window, and no other command can be executed until that job ends. Therefore, jobs should be run in the foreground only if they are interactive or very short. It is possible, however, to move a job from the foreground to the background. To do so first type **^z** (hold down the "control" key and then press the "z" key), this action suspends the job and brings you back to the "prompt", and then type **bg** to continue its execution in the background. Typing **fg** will bring the job back to the foreground.

&

A more effective way is to type the **&** symbol at the end of the command line which executes the job. This symbol tells the system to run the job in the background (without first running in the foreground). Example: **charmm** < *input* **&**.

jobs

Each job is assigned a number, which can be used to address the job. To see the list of active jobs in your UNIX window type **jobs**. Each active job will appear with a serial number and information about its execution status. This information is useful if you have more than one job in the background. For instance, if you have two jobs and you want to bring the first one back to the foreground, type: **fg %**1 (where the number following the • symbol indicates its number in the **jobs** list).

ps

The **ps** command will give a list of all your running processes (these include your jobs, but also other processes such as those that control the window operation). Each process is assigned an ID number (PID), and the **ps** command shows among other things how long each job has been running. The **ps** PID numbers can be used instead of the **jobs** % numbers.

top

Use the **top** command to display and update information about the "top cpu processes", namely the jobs that take up most of the computer's CPU time (active jobs)

kill

Another important job control operation is the one that allows you to terminate a job before its execution is completed (e.g., if the job was submitted with the wrong input). The **kill** command does the trick. This command has many options but its most useful format is: **kill -9** <job ID>. Example: **kill -9 %**1 (terminates job number 1 in the **jobs** list) or **kill -9** 13455 (where 13455 is the job PID number from the **ps** list).

nice

Finally, at times you may want to change the execution priority of your job; e.g., to reduce its priority and enable another job to take precedence on the computer CPU. To do that you may adjust the **nice** value that is associated with your job. The **nice** values range from 0 (heist priority) to 20 (lowest priority). To change the **nice** value of your job use

renice

the **renice** command. Example: **renice** 20 13455 (change the execution priority level of job PID 13455 to the lowest possible priority level).

SUMMARY

Summary of job control commands:

^z	suspend a job which runs in the foreground.
<command> **&**	submit the job for background execution (appended to the execution command line)
bg	move a job to the background.
fg <job>	bring a job back to the foreground.
jobs	list of active jobs in this window.
kill -9 <job>	terminate a job
ps	list of active processes.

top   display and update information about the "top cpu processes",

renice \<value\> \<job\>   change job execution priority to a **nice** value between 0 and 20.

\<job\>   job identifier, either the job ID (from the **ps** command) or a **%** sign followed by the job's number in the **jobs** list.

EXTRACTING INFORMATION FROM FILES

During the labs you will often generate large output files and then have to extract specific pieces of information from these files. A useful UNIX command for this purpose is **grep**, which extracts from a given file all the lines that contain a specific string; e.g.,

grep 'CHARMM' *charmm.out*

will print all the lines in the *charmm.out* file that contain the string 'CHARMM'.

The **grep** command is case sensitive. To override this and perform a case insensitive **grep** command use the **-i** option. For example, the command

grep -i 'CHARMM' *charmm.out*

will print all lines that contain the strings 'CHARMM', 'CHARMm', 'Charmm', 'charmm' etc.

In the previous example the output, in this case of the **grep** command, was displayed on the screen. As discussed above, it is possible to redirect the output of any command either to a new file (the < and > signs) or to be an input of the next command (using the pipe sign |). These options are often useful with the **grep** command.

Examples:

grep 'CHARMM' *charmm.out* > *new_file*

            writes the output of the **grep** command to the file *new_file*.

charmm < *input_file* | **grep -i** 'dynamics'

run **charmm** with input from *input_file*, and display only those output lines that include the string 'dynamics' (case insensitive).

charmm < *input_file* | **grep** -i 'dynamics' > *dyna_out*

same as the previous example, the lines that were "grepped" are saved to the file *dyna_out*.

In some cases a more powerful tool will be needed to extract the relevant information from an output file. In these cases you will be **awk** supplied with **awk** script files. **Awk** is a UNIX companion scripting language for manipulating data. Detailed instructions how to use these script files will be given when needed. In general the format will be

awk -f *file.awk* *my_file* > *output_file*

where *file.awk* is an **awk** command file (will be supplied in this *Guide* whenever needed), *my_file* is the file to be manipulated and the *output_file* is where the results will be saved.

ADDITIONAL INFORMATION

Additional information on all UNIX commands can be obtained from the **man** on-line manual. Type **man** *command-name* to get detailed listing of the command and its different options (e.g., **man** ls). If you do not know the name of the command you are interested in but have some clue to its function you can first search by keywords (**man** -k) to locate the desired command.

I. INTRODUCTION

CHARMM stands for **C**hemistry at **Har**vard **M**acromolecular Mechanics. This is a general and flexible software application for studying the structure, energetics, and properties of molecular systems. Systems can range in size from small organic molecules to large protein complexes embedded in a solvent or phospholipid membranes. Using an empirical energy function it is possible to use CHARMM for performing energy minimizations, dynamics simulations, vibrational analysis and thermodynamics calculations. The computational results can be comprehensively interpreted using numerous CHARMM analysis tools. CHARMM has its own command language, which enables the user to compare structures, evaluate conformational energies, calculate correlation functions, analyze dynamic trajectories and more.

The user interacts with CHARMM through a command interface. Often this is accessed through *CHARMM input scripts*, but CHARMM can also be used interactively from within visualization programs, such as InsightII or QUANTA. CHARMM uses units of Å for length and kcal/mol for energy. Time is usually given to in picoseconds.

NOTE: CHARMM is not the only molecular mechanics program available. Other well known molecular mechanics programs are Discover and AMBER. Despite the differences between these programs they have many features and properties in common. Thus, while the computational studies described in this *Guide* use CHARMM, they can be adapted to other molecular mechanics programs in a straightforward way.

II. DATA STRUCTURE

To represent large biomolecules on the computer CHARMM uses a detailed data structure that includes diverse information about the molecule: composition, chemical connectivity, atomic properties, internal coordinates, force constants, energy parameters and more. This information is contained in various library files and is put together by the user to represent the specific molecule of interest.

Following is a brief description of the main files involved in representing a molecule on the computer. You will use these file when constructing the CHARMM job necessary for Lab 4.

RTF

1. RTF   = Residue Topology File
    This library file specifies how atoms are connected to form amino acids and other small molecules (which will serve as building blocks for larger molecules). It also contains definitions of all atom types and of the partial charges associated with each atom in different molecular building blocks.

PARAM

2. PARAM = Parameter File
    Associated with the RTF is the PARAMeter file which contains the necessary parameters for calculating the energy of the system. This library file contains force constants and equilibrium distances for bond stretching, angle bending and dihedral motion as well as van der Waals parameters for a Lennard-Jones 6-12 potential (see the ENERGY Primer given earlier in this Lab Manual for details on the energy function).

PSF

3. PSF   = Protein Structure File
    Specific molecules are described in the form of a protein structure file (PSF). This file, which is generated using the information contained in the RTF, contains the detailed composition and connectivity of the system of interest (The PSF bares much resemblance to the MSF file generated by QUANTA). In general, the PSF can even include several molecular entities, defined as segments, which can range from a single macromolecular chain to a collection of water molecules.
    The PSF must be specified before any calculation can be performed on the molecule. Since the PSF constitutes the molecular topology but does not contain information regarding bond lengths, angles etc. it is necessary to add the parameter and coordinate information for any computation to take place.

CRD

4. CRD   = Coordinate File
    CHARMM uses Cartesian coordinates as its primary representation system, although internal coordinates (bond lengths, bond angles and dihedral angels) can also be used. Cartesian coordinates are usually obtained from crystal or NMR structures (from the Protein Data Bank). Alternatively, they can be generated from the values detailed in the RTF put together with some secondary structure information.

In those cases where coordinates are not available for some or all of the atoms in the structure (e.g., hydrogen atoms in crystal structure) they may be automatically added by CHARMM.

5. IC Table = Internal Coordinate Table

IC

The IC Table contains the bond lengths, angles and dihedrals etc. for our system. The IC Table is simply a list of relevant bond, bond angles, dihedral angles and improper dihedrals information, which is originally contained in the RTF. This table can be generated when using the SETUP keyword while generating a PSF.

The IC table is helpful when interpreting the structure of a molecule. In order to use it one has to be familiar with the notation used in this table. For any 4 connected atoms, i, j, k, and l, this table gives you: the i-j bond length, R(IJ); the i-j-k bond angle T(IJK); the dihedral angle i-j-k-l, PHI(IJKL); the other bond angle T(JKL); and the far, k-l, bond length R(KL).

The table also includes improper dihedrals, marked with a star. These are coordinates that are handy for keeping sp2 atoms planar and sp^3 atoms in a tetrahedral geometry. The atom marked with a star defines the center of the improper dihedral. The somewhat formidable notation at the top of the IC table can now be better understood. If there is a star on an atom, then PHI refers to an improper dihedral (rather than a proper one) and we use the second set of arguments for the bond lengths and angles. That is, R(I(J/K)) means that for proper dihedrals we are looking at R(IJ), whereas for improper dihedrals we are looking at R(IK). Same for the angles, T(I(JK/KL)) means that proper dihedrals this number is for T(IJK) and improper dihedrals mean we are looking at T(IKL). You will see an example of this table in Lab 4.

III. GENERAL RULES in CHARMM

CHARMM commands can be executed interactively (from within QUANTA or directly from the UNIX shell). However, to save typing (and miss typing) when doing repetitive calculations it is more efficient to put all the CHARMM commands in a CHARMM command file and submitting this file to the computer as a batch job. By simply modifying a few command lines in the input file one can easily do a number of repetitive calculations with little excess typing. The convention is that CHARMM command files are suffixed by *.inp*, e.g., *charmm.inp*.

The specification of a commands file is governed by a few rules:

1. The command line:
    The first word in every command line defines the command. The command line is written in free format and is generally echoed to the standard output. Various options and operands, which are specified by keywords, generally follow the command. In most cases CHARMM requires only the first four letters of the keyword (the rest of the word is superfluous and can be omitted). For clarity we will capitalize the first four characters of each command.

    Examples: OPEN READ FORMatted UNIT 27 NAME *RGDS.CRD*
                   COORdinate ORIEnt RMS

2. Titles:
    Titles can be provided whenever file manipulation is performed (i.e., reading or writing). Command and stream files should begin with an informative title, which is then followed by any number of commands. All title lines begin with a * followed by the title text. A line containing a single * terminates a title.

* Title
*

    Example: * First Biophysics job
                   *

3. Variables:
    CHARMM excepts user-defined variables that can be modified during execution. These variables are often used for specifying file names, loop variables etc. A variable name consists of a single alphanumerical character. A value can be introduced to a variable by the SET command, and be retrieved later using the prefix @. The

SET var

@var

type of the variable (e.g., character or integer) is determined according to the type of value you assign to it.

Example: SET file *lab5/lib/toph19.inp*
             OPEN READ FORMatted UNIT 25 NAME @file

?var

In addition to user-defined variables there are also internal CHARMM variables, which are assigned values as specific commands are executed. The value of these variables is retrieved using a question mark as a prefix. For example the SELEct command returns the number of selected atoms in the CHARMM variable ?NSEL.

4.   Stream files
In regular programming subroutines are often used to maintain the program clear and simple. This is also true for CHARMM input files. It is often more convenient to include portions of the main command file in CHARMM "subroutine" files, named *stream files*, and call them when necessary.
      A stream file starts with a title, includes as many CHARMM command lined as you like, and ends with the command RETUrn. The convention is that *stream files* are suffixed by *.str*; e.g., *load_ psf.str*.

STREam

One activates the stream file using the STREam command:

Example: STREam *load_ psf.str*

5.   Comments:
Comments can be entered on new lines anywhere in the command file, or at the end of any command line. A comment is indicated by an exclamation mark (!) at its beginning.

! comment

NOTE   CHARMM has extensive internal error checking, so if you mistype a command or operand, CHARMM will provide you with suitable (and not so subtle) error message.

IV. RUNNING CHARMM

CHARMM runs on many of platforms ranging from super computers and workstations running UNIX to PC's running LINUX. On most computers the usual mode of operation will be through a batch command file, although on small machines (such as Silicon Graphics workstations) an interactive mode is also possible.

1. Interactive Mode

    In this mode of operation, which in not suggested for large calculations, you enter the command lines one by one, pressing the RETURN key at the end of each line. If you ever get into trouble, such as getting lost or hopelessly confused, with CHARMM , hold the CTRL key down and type c. This action will terminate CHARMM; you will then have to run CHARMM again and repeat everything up to that final mistake.

    You can use CHARMM directly by typing in lower case

    charmm

    You should then get a message indicating which version of CHARMM you are running, the current time and date and your name. Continue by typing in the titles and the different commands. CHARMM will direct its output to the screen.

2. Batch Mode

    When running CHARMM in a batch mode you have to put all the CHARMM commands in a file named, for instance *charmm.inp*. Use whatever editor you like to write and edit this input file (**jot, vi, emacs**).

    This file should be directed into CHARMM by typing

    charmm < charmm.inp > charmm.out

    The output, which was previously directed to the screen, is now directed to the file *charmm.out*. The results can be viewed by editing this output file using any editor, or by displaying its content on the screen. Use **cat** to display the whole file or **more** to display it page by page (use the space bar to get the next page, type **q** to quit).

V.   A SAMPLE CHARMM COMMAND FILE (*charmm.inp*)

The following is the CHARMM command file you will construct in Lab 4. More details are given in the instruction to that lab, it is copied here just as an example of a complete CHARMM job.

```
* First Biophysics job                        ! Job title
*

OPEN READ CARD UNIT 25 NAME lib/toph19.inp  ! Read RTF

READ RTF CARD UNIT 25
OPEN READ CARD UNIT 26 NAME lib/param19.inp ! Read Param file

READ PARAmeter CARD UNIT 26


READ SEQUence CARD                            ! Read sequence
* RGDS
*
4                                             ! number of residues
ARG GLY ASP SER                               ! the sequence
GENErate RGDS SETUp                           ! Generate PSF and
                                              ! setup IC Table


! Read in the coordinates
OPEN READ FORMatted  UNIT 27 NAME RGDS.CRD
READ COORdinate CARD UNIT 27


! Specify treatment of non-bonded interactions_
NBONd CUTNb 150.0 CTONnb 145.0 CTOFnb 149.0 SHIFt -
      VSWItch CDIElectric EPSilon 1.0

ENERgy                                        ! Calculate energy

IC FILL                                       ! Fill IC Table
PRINt IC                                      ! Print IC Table

STOP
```

VI. DOCUMENTATION

CHARMM is described in the following paper:

Brooks B.R., Bruccoleri R.E., Olafson B.D., States D.J., Swaminathan S. and Karplus M., "CHARMM: A program for Macromolecular Energy, Minimization, and Dynamics Calculations", *J. Comput. Chem.* (1983), **4**:187-217.

Additional detailed documentation can be found in the "doc" filed within the CHARMM distribution directory. This information is also available on-line from Accelrys Inc. Printed documentation can also be obtained from Accelrys.

Mastering the intricate details of a molecular mechanics simulation package, such as CHARMM, is beyond the scope of this *Guide*. Nonetheless, it is important that when performing molecular simulations you will be familiar with the computational processes that are being carried out. To facilitate this goal we have prepared CHARMM *template files* to assist with the learning process.

Throughout this *Guide* you will be given CHARMM *template* files, each designed for the different computational procedures. These files, which are indicated by a *.tmpl* suffix, are located in the *Guide* directories provided on the CD that accompanies the *Guide*. You are expected to understand these command files, and then modify them according to the instructions.

The following directions should be followed whenever a CHARMM *.tmpl* file has to be modified:

i) First change the name of the *xxx.tmpl* template file to a *xxx.inp* input file,

   mv *xxx.tmpl xxx.inp*

ii) Carefully read the content of the template file and make sure you understand the different commands that are used. Refer to the CHARMM Primer for assistance. You may use a CHARMM Manual as a reference source. To view the content of the file use either an editor (e.g., **jot**, **vi** or **emacs**) or display it on the screen (**more** or **cat**).

iii) Identify what modifications you are required to introduce. These are flagged by the word MODIFY and by the **???** symbol. The modifications typically involve specifying file names and modifying one or two command lines.

iv) Introduce the required modifications.

I. OBJECTIVE

The three-dimensional (3D) structure is an important characteristic of any molecule. Understanding these 3D structures becomes even more important when dealing with biomolecules, proteins in particular, where function is directly related to structure. To view these 3D structures we use molecular visualization packages, such as Insight II or QUANTA.

In this first lab you will become acquainted with a molecular visualization program (QUANTA) and learn some of its basic operations. In particular, you will learn how to move and rotate the molecule, to select colors, to choose atoms for display, and to modify the molecule. You will also learn to import files with other formats and to use the Builders.

NOTE: Many other molecular visualization programs are also available, for example: InsightII, Sybyl, VMD and RasMol. This lab can be easily adapted to be used with other molecular visualization programs

II. BACKGROUND

MODEL

Simulating a molecule on the computer requires an abstract representation of that molecule in the computer's memory. This abstract representation is called a MOLECULAR MODEL and is a prerequisite for any type of molecular computation. More than one molecular model can be constructed for a given molecule and on the other hand a single type of model can be applied to many molecules.

ATOM

The fundamental unit of the molecular models you are going to use in this course is the ATOM. An atom is represented by a charged point mass with no directional properties and without internal degrees of freedom. The atoms are connected by virtue of the bond interaction energies. This means that the bonds themselves are not independent entities in the model, and their characteristics (such as a single bond vs. a double bond) are introduced only through the properties of the interacting atoms.

There are several ways to construct an ATOM-BASED MOLECULAR MODEL on the computer. Three very common models, which are recognized by CHARMM, are:

19

i)   All Atom Model - in which all the atoms of the molecule are represented in the model.

ii)   Extended Atom Representation - in which all hydrogen atoms are combined with neighboring heavy atoms to which they are bound; i.e., instead of representing a CH_3 group by four atoms (as in the all atom model) they are represented by a single 'extended atom' or 'united atom' of mass 15 a.u.

iii)   Polar Hydrogen Model - in which polar hydrogens (e.g., hydroxyl hydrogens) are explicitly included in the model, while all other non-polar hydrogens (such as CH_2 hydrogens) are embedded in 'extended atom' as described above. This model is a refinement of the 'extended atom' representation.

The 'extended atom' and the 'polar hydrogen' models were originally introduced to reduce the effective size of the system in order to overcome computer power limitations. These models are rarely used anymore.

III. PROCEDURE

By now you should have a personal account and a personal password on a UNIX machines. Before you start it is essential that you become acquainted with the UNIX operating system.

Read the INTRODUCTION to UNIX before you begin

• Log in to the computer. You will be in your "home" directory.

• Open a new window by clicking on the **UNIX shell** option in the Tools box, which you can find at the upper left corner of the screen (on Silicon Graphics workstations; this may vary on other workstations). Make sure the mouse is in this window before you type in any command. Iconize the WorkSpace window by clicking on the small square at the upper-right corner of the gray title bar. You can get the window back by clicking on the icon with the right mouse-button and choosing **Restore** (or double clicking on it with the left mouse-button).

- Make a new sub-directory called *lab1* (**mkdir** *lab1*) and go to that directory (**cd** *lab1*) [watch upper/lower case].

- Copy the content of the *$Lab/lab1* directory to your sub-directory by typing:
  cp *$Lab/lab1/** .   (Note: the dot is part of the command)

- Type **ls** to view the content of the directory.

A. STARTING THE QUANTA SESSION

QUANTA

MSF format

QUANTA uses molecular structure files (MSF; indicated by the extension *.msf*) to manipulate molecules. An MSF file contains a list of all the atoms in the structure, their 3D coordinates, the connectivity and information about atom types, atom numbers, and charges. Structures can be read into QUANTA from a MSF file. If the molecule you are interested in has no MSF file, other types of files (with atomic coordinates and additional information) can be read in and then converted to MSF format (e.g. protein databank files). Finally, structures can be created from scratch in QUANTA and written to a new MSF file. In this lab you will practice all three techniques.

Once you have your molecule set up in QUANTA, you will probably want to look at it, calculate distances, evaluate angles and more. As proteins can be quite large, a clear and informative display of the protein is important. This is done by applying colors to different segments of the molecule and/or by displaying only chosen parts of the structure.

Step by step instructions will be given for this session.

Start up QUANTA by typing the command *quanta*. The system informs you that the necessary startup files are not present. Click on **YES** with the left mouse button to have the files created in your *lab1* sub-directory.

QUANTA will now open and five windows will appear on the screen. The different window types are:
1. The ***molecule window***: This window contains 4 areas: (i) The viewing area, which is the big black square. This is where molecules are displayed. (ii) The menu bar above the viewing area. Every item on the bar contains a group of QUANTA functions. Try moving the mouse to a menu item and keep the button pressed. You will see the

list of functions in that menu item. Throughout the labs you will get to know more and more of the functions. (iii) The command line below the viewing area, which can be used to type in commands instead of using the mouse. (iv) the message line below the command line, which displays instructions and error messages.

2. The *Textport*: This is the window below the *molecule* window. It displays messages and information about the displayed structure.

3. The *Molecular Management* window: This window, which partially covers the *Textport* window, controls the display. It lists all open molecules, their size and their current status; i.e., whether they are currently displayed (**Visible**) and whether they enter the calculations (**Active**). The status of a molecule can be changed by clicking on the appropriate **YES/NO** indicator. Clicking on the molecule's name or size will display relevant information in the *Textport*.

4. The *palettes*: In the upper right corner you will see two palette windows **Geometry** and **Modeling**. They contain functions that can be performed by clicking with the mouse on the desired line. Many more palettes exist, and are displayed when you pick functions from the menu bar.

5. The *dials*: In the lower right corner. By operating these dials with the mouse you can rotate and translate molecules, scale them, and change color definitions.

- You are now ready to read in a structure. Click with the left mouse button on the **File** pull-down menu, and while keeping the button pressed, move down to the **Open** function to open the *.msf* file. Release the button. You will now get a window with all MSF files in your current directory displayed. You can also move to other directories. Try **Dir Up**, and you will see the contents of your home directory. There are probably no *.msf* files there. Every sub-directory is shown by its name followed by a /. You can move to the sub-directories by clicking with the mouse on the desired one. Later on, when you will have many sub-directories and many files, you will be able to read in any file this way.

- Now go back to the *lab1* sub-directory, select the structure *PEPTIDE.msf* and **open** it.

B. MOLECULAR DISPLAY

- An α-helical peptide of 20 amino acids is displayed in a so-called 'stick' model. There is no distinction between single, double, and triple bonds. Atoms are colored according to atom type (C: green, N: blue, O: red, H: white). NOTE: If you see only a small part of the molecule click on **Reset View** from the *dials*. The molecule will be centered and fitted into the window.

- Try out some other display styles:

RAY TRACE

Select the **Ray Trace** function from the **Draw** menu item. The peptide will now be drawn with more realistic atom sizes. This drawing may take a while. It cannot be rotated. Click in the *molecule* window to get back to the stick drawing.

SOLID MODEL

VdW MODEL

A similar picture can be obtained by selecting **Solid Models** from the **Draw** menu, and then selecting **Van der Waal's**. This drawing can be rotated. Place the arrow in the solid model window and move it left to right and up and down, while pressing the middle mouse button. You can rotate the structure in any direction you want. We will get back to the rotation function in a short while. Note, when you chose the Van der Waal's option an *Object Management* window appears at the bottom right of the screen. To get back to the 'stick' model toggle the **No** in the **Displayed** column into **Yes** by clicking on it. The *Object Management* window will disappear and the 'stick' drawing restored. Click on the **Reset View** dial if necessary.

- Many QUANTA functions ask you to pick one or more atoms. This is done by pointing the mouse arrow at an atom and clicking on the left button. If you are not executing a function, only the label of that atom will appear on the screen. This label comprises of the atom's name, the segment name (in this case there is only SEG1) and its residue number, e.g., **C_SEG1:10**. Additional information about this atom, which includes the name of the MSF file and the residue type, is shown in the *Textport*. Try picking a few atoms.

VIEW

To remove all atom labels, pop the *Geometry* palette by clicking the left mouse button in the gray title bar at the top of the *Geometry* palette. Now pick **Clear ID**, and all labels are removed. (Labels can be removed by picking the same atom twice if **Repick Deletes** is selected from the **ID Mode and Style** option in the **Preferences** menu.)

- To rotate the molecule around the z-axis (axes are indicated in upper left corner) press and hold the left mouse button on **Z-rotate** in the **Dials** menu. When you release the button, the rotation stops. The further to the right in **Z-rotate** you press the mouse, the faster the molecule rotates clockwise. The further to the left, the faster it rotates counter-clockwise. The same goes for the other **rotate** and **translate** dials.

  An even easier way to rotate the molecule is by using the mouse. Putting the mouse in the *molecule* window and, while pressing the middle button, move it up/down or left/right, to rotate the molecule around the x-axis or y-axis, respectively. Rotation around the z-axis is done by pressing the right button and moving left/right. The molecule is returned to its original orientation by clicking on the **Reset View** dial.

- The molecule can be scaled up and down with the **Scale** dial. The **Clip** dial determines how far "into the screen" atoms are still displayed. If you decrease the Clip width, some atoms that lie in the back will not be displayed anymore. Decreasing it even further will cause the molecule to disappear. This function enables you to view only a slice of a molecule. This is often useful when viewing large protein structures. Click on **Reset View** to get the original view.

- Pop the *Geometry* palette in front of the *Modeling* palette. The palette contains several functions, of which a few are already highlighted. Whenever you select a function, it is checked and highlighted. Some functions automatically turn off after you have completed them. Others, however, need to be turned off by selecting them a second time.

DISTANCE &
ANGLE
MONITORS

- Select the **Distance** function. You can now calculate distances between atoms by picking 2 atoms. A distance monitor (the dashed line) is also displayed, as this is highlighted in the palette. If you turn **Show Distance Monitors** off, the dashed lines will not appear, but you can still measure distances (reading the results from the *Textport*). In the same way, you can measure angles or dihedral angles between atoms, by selecting **Bond Angle** or **Dihedral**. You will have to pick 3 or 4 atoms respectively. Select **Distance** and **Continuous Pick Mode**. If you now pick 2 atoms, the distance between them is displayed. Pick a third, and the distance between the second and third is displayed, etc. the same can be done with angles and dihedrals. Now turn off any **Distance**, **Bond Angle** or **Dihedral**

tool that is still highlighted, and click on the **Delete Monitors** tools
to clear the display.

> You can now begin to answer the LAB WRITE-UP questions.
> Additional questions are scattered throughout the lab

Q1:   Residues 1-3 of the structure PEPTIDE are in a so-called
extended conformation, whereas residues 5-16 are folded
into an α-helix. These different forms are almost exclusively
determined by the Φ and Ψ (phi and psi) dihedral angles.

The definitions of the different dihedral angles are (where i
indicates residue number):

Φ_i : $C_{i-1} - N_i - CA_i - C_i$
Ψ_i: $N_i - CA_i - C_i - N_{i+1}$
$\omega_{i/i+1}$: $CA_i - C_i - N_{i+1} - CA_{i+1}$

The ω dihedral defines the torsion of the peptide bond
between residues i and $i+1$.

Calculate the Φ, Ψ, and ω dihedrals of residues 1 to 3, and
residues 7 to 10. (NOTE: Φ_1 is undefined). Does the ω
dihedral vary a lot? Why/why not?

HYDROGEN BONDS

• To see the hydrogen bonds that stabilize the helix select the
Hydrogen Bonds option from the *Modeling* palette. Click again on
the **Hydrogen Bonds** option to remove the hydrogen bonds
indicators from the display.

Q2:   α-helices are stabilized by hydrogen bonds between
backbone carbonyl oxygens of residue i, and backbone NH
of residue i+4. Stable hydrogen bonds have a distance of ~
2.7-3.1 Å between the "donor" and "acceptor" heavy atoms
(in this case the N and O) and an angle "donor-H-acceptor"
(in this case N-H-O) in the range of $180° - 30°$.

Find these distances and angles for the 6 hydrogen bonds in
the region between residues 7 and 16. Namely, the hydrogen
bond between residue 7 and residue 11 (H-bond 7-11), the

hydrogen bond between residue 8 and residue 12 (H-bond 8-12) ... up to H-bond 12-16.

Where does the α-helix hydrogen bond pattern end according to these calculations?

C. DISPLAY ATOMS

Reducing the number of atoms displayed in the *molecule* window helps you focus on the details of the molecule. To do so you have to specify a "display selection". In QUANTA there are a number of preset display selections as well as a possibility for user-defined specifications, where any selection can be specified. The selection criteria are automatically stored in a display selection file (*.dsf*).

- Select **Display Atoms** from the **Draw** menu. A pull-right menu with a list of possible selections is automatically displayed. The first 5 options are preset display selections (e.g., protein backbone). The last option (**Selection Tools**) will be discussed below.

BACKBONE

- From the pull-right menu select **Protein Backbone**. Only the backbone atoms of the peptide are now shown. The message in the *Textport* tells that 103 atoms will be displayed (Note, the molecule still has 201 atoms).

C_α **TRACE**

  Now display only the protein **C-Alpha Trace**. Although there are no real bonds between α-carbons, the program will connect them to show the general shape of the backbone. This is a very useful option for examining the structures of larger proteins. Rotate the molecule until you look down along the long axis of the helix.

  Restore the display of all the atoms by selecting the **All Atoms** option.

- Choosing **Selection Tools** opens a new palette which includes, in addition to the preset display selections, type and range criteria (e.g., residue type) and several options for interactive element picking (using the mouse). Select **Quit** or **Finish** to exit this palette.

- A useful selection criterion can often be to select all atoms that are within a given distance from a specific atom or residue. Open the **Selection Tools** from the **Display Atoms** item in the **Draw** menu. Make sure the **Include** tool is highlighted, and select **Proximity**

AROUND
ATOM

Tools .... In the new palette that appears choose the **Around Atom** option. Now pick any atom in the molecule. All atoms within the specified distance (default 5 Å) from the atom you picked are highlighted. To change this default value, choose **Set Radius/Length** and enter a new value, select **Clear** and pick the central atom again. Select **Exit Proximity** and then **Finish**. Only the marked atoms will be displayed.

RESIDUE
RANGE

- The display selection can be further refined by the **Exclude** tool. Suppose you want to display all the atoms except those of residues 1 and 6. Choose **Selection Tools** from under **Display Atoms** in the **Draw** menu. Make sure the **Include** tool is highlighted, and select the **All Atoms** option. All the atoms are highlighted in red. Now click on the **Exclude** tool and choose the **Residue Range...** option. Select the desired range and click **OK**. These residues are now de-selected. Click **Finish**, to view the resulting cut down display.

NOTE: These changes affect only the display and will not affect any calculations performed on the molecule.

- To display all atoms again, select **All Atoms** from under **Display Atoms** in the **Draw** menu. All atoms in the peptide are again displayed.

D. CHANGE THE MOLECULE

Sometimes you want to remove a part of the molecule. This means that this part will not be included in any following calculations and, of course, it will not be displayed (because it isn't there anymore). To see how this works, select the **Active Atoms** option from the **Edit** menu. The pull-right menu that appears has two predefined options (All atoms, All except Solvent) and a link to the Selection Tools. Choose the **Selection Tools** option. The palette that appears is identical to the *Display Atoms* palette.

SELECTION

- Instead of defining the selection interactively (which would work exactly the same as you did in the Display Selection before), you can also type in your selection. This is useful if you already know which residues you want to delete from the structure. Select **Type in a Selection....** You will see a dialog box that asks for selection commands and displays five empty lines. A selection is usually built stepwise as a series of commands that specify and refine the selection. For example, to remove the first 4 residues from the molecule, you can type in the following three lines (you move from one line to the next with the Tab key or the mouse):

all   (initially include all atoms)
excl   (the atoms in the next line will be excluded)
zone 1 to 4   (residues 1 to 4 should be excluded)

When done click **Done** to see your selection. Click **Finish** to exit and see that the first 4 residues are deleted.

NOTE: Instead of 201 atoms you now have only 164 atoms (see the *Textport* and *Molecular Management* window).

• To save this new molecule in a new MSF file, select **Save As** from the **File** menu. Make sure that all five options are selected and press **OK**. You are asked for a new filename. Type **newpeptide** on the highlighted line (which displays *.msf on it) and select **Save**. The new MSF file will be called *newpeptide.msf*. Remember that your old peptide with 20 residues has its own MSF file *PEPTIDE.msf*.

E. COLORS

An important method for improving the usefulness of a display is to change colors. QUANTA allows you to assign colors in any way you like.

• Select **Color Atoms** from the **Draw** menu. Again a pull-right menu appears with a few predefined options and a link to the Selection Tools. Choose **Selection Tools**. Two palettes appear. The one named *Color Atoms* looks suspiciously like the one you saw in the Display Selection. The other is named *Color Schemes and Utilities* and is partially hidden behind the *Color Atom* palette. Bring the second palette forward by clicking on its title bar. Move it slightly away from the *Color Atoms* palette. You can change the color using the **Next Color Number** tool or the **Choose Color Number...** tool. There are 14 colors in total. Colors 1 to 6 can be seen by clicking on the '5' on the bottom of the *dials* box. Colors can be altered at will by the **H, S** and **I** buttons next to the color numbers. H (hue) will change the actual color, S (saturation) controls the amount of white in the color, and I (intensity) determines the brightness of the color. Hold the left mouse button pressed on the left or right side of any of the H, S or I boxes, and you will see that the color changes. By changing color 1 or 2 you will see that the colors of the atoms also change. The color selections are automatically put in a color selection file (*.csf*). All colors can be reset to their original values by picking the **Color Definitions** function from the **Preferences** menu, and selecting the **Reset All** option. Colors 7 to 14 can be changed by picking the '**6**' at the bottom of the *dials* box. The *dials* box is put back to the rotation and translation dials by picking the '**1**'.

- Let us for example color the backbone in red and the side chains in blue. Bring forward the *Color Atoms* palette. Select **All Atoms** to color all the atoms in blue (color 2). Now choose **Next Color Number** (from the *Color Schemes and Utilities* palette) the and **Protein Backbone** (from the *Color Atoms* palette) to color the backbone in red (color 3).

- Similar to the Display Atom selection, color selections can also be typed in. Select **Type in Selection** from the *Color Atoms* palette. Try:
  all = col 4
  zone 8 to 12 = col 2
  This will color every atom yellow (color no. 4), except residues 8 to 12, which are blue (color no. 2). Click the **Done** button to see the results.

- In addition QUANTA allows you to color atoms and residues according to several physical properties: atomic type (default), atomic charge and residue polarity. These options appear in the *Color Schemes and Utilities* palette.

  COLOR BY POLARITY

  Lets try coloring the molecule according to residue polarity. Bring the *Color Schemes and Utilities* palette forward. Select the **Amino Acids by Polarity** option and then choose **All Atoms** from the *Color Atoms* palette. This will color residues according to these rules:

positive charge	= color 2	(blue)
negative charge	= color 3	(red)
uncharged polar	= color 4	(yellow)
nonpolar	= color 5	(white)

- Also try to Color by **Atomic Charge** (don't forget to choose **All Atoms** afterwards). QUANTA has predetermined charges assigned to each atom, and this option will color atoms according to them. Highly negative charges are given color 1, highly positive charges color 14, and in-between charges a color in between, depending on the actual charge (charge 0.0 will be color 7). It is clearer to make the 1 to 14 scale a smooth one with colors varying from red (1) [=negative] to blue (14) [=positive]. This can be done by selecting the **Red to Blue Smooth Range** option. You can see that hydrogens are positively charged, while oxygens are negative. Many side chain carbon atoms are close to being neutral (green). Now reset all colors to the original ones by choosing **Default Colors**.

COLOR BY CHARGE

- Finally, choose the Color by **Element Type** option. This will show atoms according to the following rules:

C: color 1   N: color 2
O: color 3   S: color 4
H: color 5   P: color 6
F/Cl: color 7   I: color 8

- Click on **Finish** in the *Color Atom* palette.

F. SEQUENCE BUILDER

- An easy way to build a peptide from scratch is the sequence builder. Closer the current MSF by choosing **Close** from the **File** menu. Now click on *newpeptide.msf*, and **OK**. Pick **Sequence Builder** from under **Builders** in the **Applications** menu. You are asked to pick a residue library, which contains information on how residues are built up. Pick the *AMINOH.RTF* and **Open** it.

- Now you can build your own peptide by clicking on the amino acid types displayed on the left. Continue doing this until you have a 12-residue peptide. You can insert a new residue in the sequence by clicking on the hyphen before the insertion position. The yellow square will now move to that position, and you can insert residues by selecting a residue from the list. Residues can be deleted by first Marking them, by clicking on a residue in the sequence. That residue name is now displayed in yellow. If you pick a second one further down the sequence, all residues in between will also be marked.

Delete the marked residues by picking the **Delete Marked Item** option from the **Edit** menu of the *Sequence Builder* window.

SECONDARY STRUCTURE

- Now define the conformation of this peptide. Select **Set Secondary Conformation** from the **Conformation** menu and mark all residues that need some secondary structure, in this case all of them. Click **OK** and select a descriptor, for instance **Right handed Alpha Helix**, and **OK**. Now you are finished with the secondary structure, so pick **Done**. All information necessary to generate this molecule is now present, and you can go back to the molecular modeling part. Select **Return to Molecular Modeling** from the **Sequence Builder** menu and pick **YES** for saving your sequence. Give this sequence file a name, and press **save**. You will now return to the *molecule* window.

- Your generated peptide is displayed, and an MSF file is already written to your directory. To clearly see the helical structure select **Protein Cartoon** from under **Solid Models** in the **Draw** menu. Use the mouse to rotate the molecule.

- To delete the cartoon go to the *Object Management* window, and change **NO** into **Yes** in the **Delete** column. Now, **Close** this MSF file.

G. IMPORT FILES & ADVANCED DISPLAYS

PDB

- The final exercise is to read in molecules from files of a different format. A well-known format is the Protein Data Bank (pdb) format. Select **Import Single Structure** from the **File** menu, and pick **Protein Data Bank** as the Import File Format. Now type *pdb8lyz.pdb* and **Import**. Pick **NO** on the question if you want to set up symmetry. The enzyme lysozyme, which cleaves glycosidic bonds in polysaccharides, is now displayed. Get a clearer view of the protein structure by using **Protein Cartoon** as in the above example. You can clearly see the α-helices. Make the display less crowded by only displaying the α-carbons, you know how.

SUGAR

- Now read in a substrate. Do another **Import**, but now read in *pdb9lyz.pdb*. The program asks if you want to keep the same dictionary, click **NO**, and select the **Sugars** dictionary (as the substrate is a polysaccharide). Do not set up symmetry, and add this molecule to your previous selection. Both molecules are now displayed, and you can see where the binding cleft of the enzyme is. For clarity color each molecule in a different color (use the **By Molecule** option from the *Color Atom* pull-right menu under **Draw**. Rotate the molecule to better your view. If you have time left, you can display some more side chains of lysozyme and look at specific interactions.

- To get an even nicer view try the following:
  - Select **Van der Waal's** from under **Solid Models** in the **Draw** menu.
  - In the *Object Management* window delete the Solid Model of *PDB8LYZ*.
  - Use **Protein Cartoon** as described above.
  - Choose **C-Alpha Trace** from the **Display Atoms** option under **Draw**.
  - Rotate the molecule to better your view.

In the *lab1* directory you will find two more *.pdb* files of interesting proteins. You can import them into QUANTA and practice your new skills.

- *pdb1mbo.pdb*: Myoglobin with O_2 bound (almost entirely α-helix). (While reading this file into QUANTA answer YES to the question "Handle alternate location as disorder"). See if you can arrange the display to get a clear view of the heme group with the bound oxygen molecule. All atoms in the heme group and the oxygen have residue type HEME.

- *pdb3fab.pdb*: An IgG Fab fragment (almost entirely β-sheet).

Q3:   Look at the lysozyme-sugar structure and at the myoglobin-oxygen structure. How do the ligands interact with their respective protein molecules? Be specific. Try to find the bonding/contact points and measure the distances between the relevant ligand/protein atoms. Try to asses the type of interactions involved (are they H-bonded? Covalently bound? otherwise?).

H. EXIT QUANTA

- You are now finished with the QUANTA introduction. Do not worry if you think you cannot remember all this. A lot of things will come back and that is the best way to learn these things.

- End your session by selecting **Exit QUANTA** from the **File** menu. Do a listing of the files in your *lab1* sub-directory (**ls**). You see that you created some new ones. If you type **quanta** again while in this sub-directory, you will find it exactly how you left it.

- Log out the computer by moving the arrow in the blue area, pressing the right mouse button and selecting **Log out**. Yes, really!

III. LAB WRITE-UP

Answer the three questions (**Q1**, **Q2**, **Q3**) that were posed during the lab exercise.

Lab 2:   Energy and Minimization

I. OBJECTIVE

In this lab you will learn how to perform advanced tasks with QUANTA. You will use ChemNote and the 3D Builder tools to generate molecules, calculate energy, setup energy minimization and perform energy minimization with constraints (distance and dihedral angle constraints).

II. BACKGROUND

To understand this lab you must be familiar with the concept of potential energy function and its realization in molecular mechanics simulation programs such as CHARMM. In addition, the concepts of minimization and constrained minimization must be understood. Below is a short discussion of these topics; a more detailed discussion can be found in the references.

Potential Energy Function

The empirical energy function is a sum of many terms. In CHARMM the potential is a sum of internal or bonding energy terms (bonds, angles and torsions), pairwise non-bonded interaction terms (van der Waals, electrostatic) and other optional term (such as constraints). The total energy is thus:

ENERGY FUNCTION

$$E = \{ E_b + E_\theta + E_\phi + E_\omega \} + \{ E_{vdW} + E_{elc} \} \; (+ \text{optional terms})$$

The functional forms for each of these terms are:

I. Bonding energy terms

Bond stretching and angle bending are modeled by harmonic (quadratic) functions:

BONDS

$$E_b = \Sigma \, k_b \, (r - r_0)^2$$

ANGLES

$$E_\theta = \Sigma \, k_\theta \, (\theta - \theta_0)^2$$

where k_b, k_θ are harmonic force constants and r_0 and θ_0 are the equilibrium values of the bond length and bond angle.

NOTE: In this potential function, bonds <u>cannot</u> dissociate. This is suitable for body temperature or room temperature, but is unphysical at elevated temperatures. To describe bond dissociation accurately, a different functional form should be introduced instead of the harmonic function For example, a fairly accurate description of chemical bonds is provided by the Morse function, which is written as $E_b = D_e (1 - \exp[-\beta(r - r_0)])^2$, where D_e is the dissociation energy.

DIHEDRAL ANGLES

The <u>dihedral angle</u> (torsion) potential is a four-atom term (A-B-C-D) based on the dihedral angle about an axis defined by the middle pair of atoms B-C (the angle is defined between the planes ABC and BCD, see figure). The functional form of this energy term is:

$$E_\phi = \Sigma \left(|k_\phi| - k_\phi \cos (n\phi) \right)$$

where k_ϕ is the energy constant and $n = 1, 2, 3, 4, 6$ is the periodicity of the torsion ϕ.

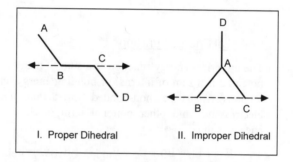

I. Proper Dihedral   II. Improper Dihedral

IMPROPER DIHEDRAL ANGLES

The <u>improper dihedral</u> (torsion) term is introduced for two reasons: (i) to maintain planarity about certain planar atoms (such as carbonyl carbon), and (ii) to maintain chirality about a tetrahedral heavy atom in the 'extended atom representation'. In addition this term provides a better force field near the minimum energy geometry, which is important for dynamic and for vibrational analysis.

These additional torsion angles are defined similar to the proper torsion angles, i.e., as the angle between the planes ABC and BCD, but in this case a different atom ordering is applied (see figure). This special interaction is modeled by a quadratic distortion potential:

$$E_\omega = \Sigma k_\omega (\omega - \omega_0)^2$$

The force constants (k) and geometric constants (r_0, θ_0, n etc.) are taken from the parameter tables (see below) based on the atom types involved.

II. Non-bonded energy terms

The two main non-bonded interactions are the van der Waals (vdW) and electrostatic interactions, which are calculated in a pairwise additive way between all non-bonded pairs (i.e., atoms that are at least four atoms away along the bonded sequence).

VdW

The vdW interaction is modeled by the Lennard-Jones function

$$E_{vdW} = \Sigma_{ij} \left(A_{ij}/r_{ij}^{12} - B_{ij}/r_{ij}^{6} \right)$$

ELECTRO-STATICS

and the electrostatic interaction is given by

$$E_{el} = \Sigma_{ij} \, q_i \, q_j \, / \, 4 \, \pi \, \varepsilon_0 \, r_{ij}$$

where r_{ij} is the distance between atom i and atom j, q_i and q_j are the atomic charges and ε_0 is the dielectric constant.

COMMENTS:

(1) In the above definition of the electrostatic interaction we included only point charge contributions neglecting higher moments of the interactions. It is possible to add these quadruple and higher moments to the energy function, but these topics are beyond the scope of this lab.

(2) In some empirical potential functions there are explicit terms for hydrogen-bond energies. In the current CHARMM all-atom representation these energetic contributions are embedded in the parametrization of the other non-bonded interactions.

Parameters

To actually calculate the energy of the molecular structure the parameters (force constants, minimum energy geometry, non-bonded interaction constants) must be specified. These constants are called the parameters of the energy function, and they differ from one atom to another depending on its environment.

The number of parameters depends on the functional form of each energy term. Bonds, for example, require two parameters: K_b - the bond stretching force constant (in kcal/Å^2), and r_0 - the reference or minimum-energy bond distance (in Å). For electrostatics, we need the two partial atomic charges (which depends on the molecular environment, and may differ from one residue to another) and the dielectric constant.

Since the interactions and the properties of a carbon atom, which is at a Cα position in a peptide, are different from the properties of a carbonyl carbon, CHARMM has different <u>types</u> of carbon, oxygen, nitrogen and hydrogen atoms do describe different environments (e.g., type C stands for a carbonyl carbon while type CT stands for an aliphatic carbon). However, once an atom type is defined the parameters are transferable, i.e., every time this type of atom is encountered, whether in the same molecule or in different molecules, the program will assign it exactly the same set of parameters.

Parameters are derived from experimental data and from high-level quantum-mechanical calculations. Sources for parametrization include:

- IR spectra (bond stretching, angle bending ...).

- X-ray crystal structures (geometry).

- Solvent simulation to match physical properties.

- Ab initio calculations (partial charges, geometry ...).

- Free energy perturbation in comparison to experimental data.

As you see, developing of a complete set of parameters requires the use of multiple data sources. The refinement of such parameter sets and their extension is a continuous on-going process.

Transferring parameters from one computer program to another is often <u>not</u> an easy task, due to differences in functional forms and in the way the parameters were derived. It can be done however.

Minimization

Given a potential energy function, it is often desirable to find minimum-energy conformations of the system. In some cases minimization is performed to relieve strain in conformations obtained experimentally. In other cases, finding a local or global energy minimum may be of prime interest, e.g., for determining the stable configuration of a peptide. As the task of locating the global minimum in macromolecules is often formidable most minimization algorithms are used only to locate local minima (and not the global minimum).

**MINIMIZ-
ATION**

The basic concept of local minimization is shown in the following figure. Minimizing the molecule at conformation X will result with the conformation associated with energy minimum α, while minimizing conformation Y results in the conformation associated with energy minimum β. Local minimization is unable to go from conformation Y to the global minimum α.

There are several optional minimization algorithms, some of which will be discussed in the next lab.

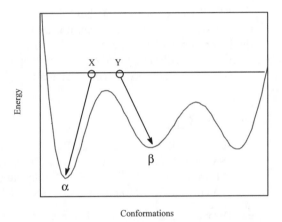

Figure 1:   A schematic one-dimensional cross-section in a potential energy surface. Shown are conformations X and Y and their corresponding local minima α and β.

Constraints

As you will see in this lab, constraints are often used to force a molecule into a desired conformation, by holding the specified variables (distances of angles) at or near their target values. It is important to understand, that these constraints are in fact additional energy terms, which act as 'penalty functions'; if they are not satisfied the energy increases significantly.

NOTE: for this reason, it is necessary to turn off the constraints
               whenever the real energy of the conformation is calculated
               (so that the artificial constraint energies would not be
               included).

Usually both distance (E_{Cr}) and dihedral ($E_{C\phi}$) constraints are
introduced using stiff harmonic interactions

DISTANCE &
DIHEDRAL $$E_{Cr} = \Sigma K_i (r_i - r_{io})^2$$
CONSTRAINTS $$E_{C\phi} = \Sigma K_i (\phi_i - \phi_{io})^2$$

where K_i are the 'force' constants (usually large values) and r_{io}, ϕ_{io} are
the target distance and dihedral values.

The harmonic constraints hold the variable near its target value but
not necessarily at that exact value. It is also possible to introduce strict (or
rigid) constraints that force the variable to be at an exact value. Formally
this means that $\delta E/\delta r = 0$ and $\delta r = 0$ for all operations. We shall introduce
such rigid constraints (SHAKE) in later labs.

REFERENCES

1.   Brooks, B.R., Bruccoleri, R.E., Olafson, B.D., States, D.J.,
     Swaminathan, S., Karplus, M., "CHARMM: A Program for
     Macromolecular   Energy,   Minimization,   and   Dynamics
     Calculations". *J. Comput. Chem.* (1983), **4**:187-217.

2.   Brooks, C.L., III, Karplus, M. & Pettitt, B.M. *Proteins: A
     Theoretical   Perspective   of   Dynamics,   Structure   and
     Thermodynamics,* Advances in Chemical Physics 71 (John Wiley &
     Sons: New York, 1988).

3.   McCammon, J.A. & Harvey, S.C. *Dynamics of Proteins and Nucleic
     Acids* (Cambridge University Press: Cambridge, 1987).

4.   Becker, O.M., MacKerell Jr., A.D., Roux, B., and Watanabe, M.
     (Eds.), *Computational Biochemistry and Biophysics* (Marcel Dekker:
     New York, 2001).

III. PROCEDURE

Create a new sub-directory *lab2* (**mkdir** *lab2*), and change directory (**cd**) to that directory. Copy all the files from the directory *$Lab/lab2* to your *lab2* directory (**cp** *$Lab/lab2/** .).

A. ENERGY PROFILE OF CYCLOHEXANE

1)   Pick the chair conformation of cyclohexane from the QUANTA/ChemNote templates and get the energy of the structure.

- Startup **Quanta**.

- Choose **2-D Sketcher** from the pull-right menu under **Builders** in the **Applications** menu. A large *ChemNote* window opens.

- Pick **Template** from under the **File** menu.

- Click on the chair conformation of cyclohexane.

- **Quit** under **File** in *Template Viewer* window.

- Pick **Paste** under **Edit**.

- **Return to Molecular Modeling** under **File**. **Save** changes as *CHAIR.mol*.

- Rotate and view the cyclohexane in the chair conformation.

Q1:   In the following sections you will be asked to <u>Record</u> the total energy and the energy components of several different cyclohexane conformations. Compile your results in a table.

- Pick **CHARMm Energy** under the *Modeling* palette to calculate the energy (it is in kcal/mol). <u>Record it</u>. You can see the contribution of different energy terms in the *Textport*.

- Pick **Minimization Options** from the **CHARMm** menu. Select Adopted-Basis Newton-Raphson minimizer.

- Pick **CHARMm minimization** from the *Modeling* palette, then recalculate **CHARMm Energy**. <u>Record</u> the final energy and its components.

- Pick **Save changes** from the *Modeling* palette and **Save to New Filename** *CHAIRm.msf*.

- To compare the minimized structure to the original one, **Reset View** (from the *dials*) and **Open** the original *CHAIR.msf* file, use the **Append** option. Now you have both structures superimposed. Color each structure in a different color (use the **Color Atoms By Molecule** option from the **Draw** menu) and rotate to view the differences. Use the *Molecular Management* window at the bottom of the screen to turn the molecules' display on and off (click on the YES/NO under **Visible**).

- **Close** the original *CHAIR.msf* file.

2) Minimize the chair conformation of cyclohexane with dihedral angle constraints to obtain the half-chair conformation, and get the energy of cyclohexane in this new conformation.

- **Reset View** on the dial. Click on the carbon atoms of cyclohexane, so that the order of labeling is clear to you.

- Go to **Constraints Options** under the **CHARMm** menu, and from its pull-right menu choose **Dihedral/Distance**. A new *Edit Constraints* palette appears.

- Now set the values you want to constrain to. Choose **Dihedral Options** from the palette and in the dialogue box enter the following information (move between fields with the Tab key):

  - Select **Constraint to A Specified Target**.

  - For the **Target Dihedral** and its **Lower** and **Upper** bounds enter = 0.00

  - For **CHARMm Force Constant** enter = 1000.00

  - Click **OK** to finish.

- Go to **Label Atoms** under the **Draw** menu, and select the **Atom Name** option to display atom names.

- Select **Define Dihedral Constraint(s)** from the new palette and with the mouse pick the four atoms defining the first dihedral to be constrained in the following order: **C5--C6--C1--C2** . Now pick the four atoms of the second dihedral to be constrained: **C6--C1--C2--C3** .

- Select **Save Constraints As** and name the file *HALFCH.con*. Pick **Exit Edit Constraints** from the palette, and choose

> **Dihedral On** from under **Constraints Options** in the **CHARMm** menu.

- Pick **CHARMm minimization** from the *Modeling* palette. You can observe the conformation of the molecule change as its energy is minimized.

- Turn **Off** the dihedral angle constraints (from within the **Constraints Options** under the **CHARMm** menu), recalculate **CHARMm Energy** and <u>Record</u> the final energy and its components. Rotate the molecule to view its structure.

- Pick **Save Changes** and **Save to New Filename** (*HALFCH.msf*).

3) Minimize the chair conformation of cyclohexane with dihedral angle and distance constraints to obtain the boat conformation, and get the energy of this new conformation.

- Choose **Open** under **File** menu. Select *CHAIRm.msf* and **Replace**.

- Pick **Dihedral/Distance** from under **Constraints Options** in the **CHARMm** menu (in the dialogue box choose **Start New Constraints Database**). Apply the following dihedral constraints as described in the previous section. Before defining <u>each</u> angle remember to set the **Dihedral Options** according to the following specifications:

	Optimal angle	Force constant
C5--C6--C1--C2	51.090	1000.00
C6--C1--C2--C3	-51.090	1000.00

- Now we want to add a distance constraint of 1.83 Å between **H8** and **H13**. First, click on the hydrogens to identify the appropriate atoms. Choose **Distance Options** and select **User Defined** in the dialogue box. Next, enter **Target Distance** = 1.83 (enter the same value also for both upper and lower limits) and **CHARMm Force Constant** = 100 (both KMIN and KMAX). Click **OK**. Now select **Define Distance Constraints** and with the mouse pick the two atoms to be constrained (**H8** and **H13**), i.e.,

	Optimal distance	Force constant
H8 to **H13**	1.83	100.00

- **Save Constraints As** *BOAT.con* and **Exit Edit Constraints**.

- To activate the constraints select **Dihedral On** and **Distance On** (click **OK** to the question that appears) from the **Constraint Option** menu under the **CHARMm** menu.

- Select **Minimization options** from the **CHARMm** menu and change the number of minimization steps from 50 to 150. Now to perform the minimization select **CHARMm Minimization** from the *Modeling* palette.

- Turn **Off** both dihedral and distance constraints, change the number of minimization steps from 150 back to 50, and minimize to obtain the best boat conformation. Record the final energy and its components and **Save Changes** from the palette. Name with a New Filename *BOAT.msf*.

4) Obtain the TWIST BOAT conformation of cyclohexane starting from the BOAT conformation you just obtained (*BOAT.msf*) set the following constraints (Start New Constraints Database):

Constrained dihedrals:

C5--C6--C1--C2	51.090	1000.00
C6--C1--C2--C3	-51.090	1000.00

Constrained distances:

H8 to **H13**	2.67	100.00
H12 to **H17**	2.67	100.00

- First minimize for 50 steps with the appropriate dihedral and distance constraints turned **On**. Then turn **Off** both constraints and minimize an additional 150 steps to get the energy of this new conformation. Calculate the final energy an Record it.

5) Record the final energy and **Save changes** to new filename
    TWBOAT.msf and then **Close** from the **File** menu.

Q2:   Based on the table you compiled in **Q1**, draw a potential
               energy diagram for cyclohexane. Which is the lower energy
               conformation?

Q3:   Compare the diagram drawn in **Q2** to the potential energy
               diagram for cyclohexane found in Morrison and Boyd's
               Organic Chemistry (3rd Edition, 1973, p.297), which is
               reproduced below. What are the possible errors in the
               diagrams you have drawn?

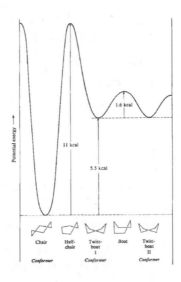

Figure 2:   Potential energy diagram for cyclohexane (adapted from
               Morrison and Boyd, *Organic Chemistry* (3rd Edition, 1973).

B. ENERGIES OF α-GLUCOSE

Q4:   In the following sections you will be asked to <u>Record</u> the
               total energy and the energy components of two different α-
               glucose conformations. Compile your results in a table.

1) **Import** the protein data bank (pdb) format coordinate file for α-glucose (*AGLUC.pdb* in the *lab2* directory). Pick the **Sugars** dictionary, and get the energy for α-glucose in this CHAIR conformation. <u>Record</u> the contribution of different energy terms to the overall energy (in the *Textport*). [in case of missing parameters just **quit** the dialogue box to setup the defaults].

2) Minimize the chair (starting) conformation of α-glucose with dihedral angle and distance constraints to obtain the TWIST BOAT conformation.

   Constrained dihedral:

O5--C5--C4--C3	57.490	1000.00
C5--C4--C3--C2	-53.290	1000.00

   Constrained distance:

O4 to O1	2.67	Scale = 100.00
C6 to H2	2.67	

   Turn on the constraints and minimize 100 steps. Turn off the constraints and get the energy of this new conformation. <u>Record</u> the contribution of different energy terms to the overall energy (in the *Textport*).

3) **Save changes** to new filename *GLUCTW.msf*.

4) **Export** the coordinates of the twist boat conformation of α-glucose as a pdb format file (*GLUCTW.pdb*). **Close** the *.msf* file.

C. D-GLUCOSE IN THE LINEAR FORM

The enzyme Xylose Isomerase performs the isomerization reaction of glucose to fructose on the linear open form of D-glucose, rather than on the cyclic form. You will generate the linear structure and fully minimize its energy (record the final energy). This was the conformation of D-glucose initially used to fit the electron density in the refinement of the x-ray structure of the xylose isomerase/glucose complex (A. Lavie, K.N. Allen, G.A. Petsko, and D. Ringe *Biochemistry* (1994), **33**:5469–5480).

The structure and chirality of D-glucose is given in the boxed diagram below. Use it as a guide to build your own D-glucose.

D-GLUCOSE

1) Use the 3D Builder to generate the linear tructure of D-glucose and then fully minimize it.

- Choose **3-D Builder** under **Builders** in the **Applications** menu.

- Choose **Display Fragment Table** to show the list of molecular fragments (clicking on TOGGLE ICONS toggles between the name of the fragment and its molecular sketch). Select the **Functional Groups** category.

- Build the D-glucose backbone from molecular fragments. First select the *ethyl* fragment twice (you will see the two fragments in the display area). Then choose **Merge Fragments** from the palette and use the mouse to pick the two bonds that should be merged. You now have a butane fragment. Click on **Reset View** in the *dials* window. Select a third *ethyl* fragment from the table and merge it to the existing butane fragment to get a hexane. Click on **Reset View** in the *dials* window. Hide the fragment table by clicking on **Display Fragment Table** again.

- Now we have to convert the hexane into a sugar. First lets change the appropriate hydrogens into hydroxyl groups. Select the **Add Atoms** tool and from the new palette that appears choose **Convert to Hydroxyl**. With the mouse pick each hydrogen that is to be converted and see how it changes to an OH group.

- To change the first carbon to a carbonyl first select **Delete Atoms** and pick the three hydrogens that are connected to it (this will remove them). Then select **Add Atoms** and choose **Convert to**

Carbonyl. With the mouse pick the first carbon atom and see it turn into a carbonyl group.

- Now you have all the atoms in place, and are ready to specify the chirality. First select **Edit Atoms** and from the new palette that appears choose **Report Chirality**. Use the mouse to pick carbons 2 through 5, the chirality is reported in the *Textport*. **Exit Edit Atoms**.

- To correct the chirality of carbons with an incorrect chirality (see figure) choose **Swap Bonds** from the *Molecular Editor* palette and pick with the mouse the two bonds to swap (the C-H and the C-O bonds). Repeat this procedure for each carbon atom which chirality is to be changed. Go back to **Edit Atoms** and re-choose **Report Chirality** to check that the carbon atoms have the correct chirality.

- To wrap things up rename the Carbon atoms. Select **Edit Atoms** and choose **Rename Atoms**. With the mouse pick the six carbon atoms and rename them C1 through C6 according to the convention shown in the above figure (the carbonyl carbon is C1). You can do the same with the Oxygen atoms.

- Congratulations! The molecular model of D-glucose is complete.

- Choose **Save and Exit**. In the dialogue box pick the "Generate Topology File - As a single residue" option, in addition to the pre-set options. Click **OK**. In the 'Adjust Total Charge' dialogue box just click **OK**, and save the molecule as *DGLUC.msf*.

- Now do a CHARMM energy minimization similar to the one you did for cyclohexane. Pick **CHARMm Minimization** from the *Modeling* palette, after setting the **Minimization Options** to 400 steps.

- Select **Save Changes** and **Overwrite** *DGLUC.msf*.

2) **Import** the coordinates of the structure of xylose isomerase + D-glucose (*wtxi.pdb* in the *lab2* directory). Examine the glucose in the binding site, select only the glucose atoms, and get the energy of glucose in the conformation in which it is bound.

- **Import** the PDB file *wtxi.pdb*. Keep the sugars dictionary and use the **new molecule only**.

- Choose the **Selection Tools** from under **Color Atoms** in the **Draw** menu. Color the molecule in one color (e.g., use **Pick Molecule**) and the glucose molecules (named GLC) in another color (e.g., use **Residue Type**). Click **Finish** to exit the Color Atom palette. You see two sugar molecules buried in the enzyme.

- To make the picture clearer go to the **Selection Tools** of **Display Atoms**. From the palette **Include** the **Alpha-carbon Atoms** and select the GLC molecules using **Residue Type** These selections will display the two bound sugar molecules in the protein C_α-carbon skeleton. **Finish**.

- Zero in on the acyclic D-glucose in subunit one (residue 950). Use the **Selection Tools** of the **Active Atoms** option from the **Edit** menu (it may take some time before the new palette appears). Choose **Residue Range** from the *Active Atom* palette and enter 950. **Finish**.

3) Compare the conformations of the bound sugar molecule to the open form of glucose obtained from 3D Builder.

- Open the *DGLUC.msf* file, which is the minimized glucose generated with the 3D Builder, and **Append** it to the current file. Now you have two molecules in your *Molecule Management* window. Choose **Reset View** from the dials.

- Compare the conformations of the two sugar molecules by doing a rigid body fit. For both molecules first display **Atom Names** as labels (use **Label Atoms** option under the **Draw** menu). Select **Molecular Similarity** in the **Applications** menu. Pick **Match Atoms** in the new *Molecular Similarity* palette. Now click on **Pick equivalent atoms**. Click with the mouse on the C2 atom in one molecule and then on its equivalent C2 atom in the other molecule. A line is stretched between them to indicate the matched pair. Continue matching atoms C3 to C6. On the *Match Atoms* palette pick **End Atom Picking** and **Exit Match Atoms**. Pick **Rigid Body Fit to Target** to fit corresponding atoms of the sugar molecules on each other. **Exit molecular similarity** and **Reset View**. You can clearly see the conformational differences between the two molecules. Record the RMS deviation between the two structures (from the *Textport*).

III. LAB WRITE-UP

Answers question **Q1** through **Q4** that were posed during the lab exercise. Also answer the following two questions.

Q5:   Compare the structure of the glucose bound to the Xylose Isomerase to the structure, which you obtained after minimizing the linear glucose. Also compare with the closed form of glucose. What are the similarities and/or differences? Can you rationalize them?

Q6:   What types of interactions are important in the binding of glucose to Xylose Isomerase? The enzyme reaction occurs on the open form of the ring, but biochemical data indicate that the closed form is initially bound. Speculate on the possible reasons for the enzyme to bind a single anomer of the closed form of the sugar. Why is the glucose molecule not bound directly in the open form?

Lab 3 :   Minimization and Analysis

I. OBJECTIVE

The objective of this lab is to gain experience with different minimization algorithms and to understand the effect of different non-bonded interaction models on the resulting structure. You will use structural tools to compare the resulting conformations. In addition, the lab intends to reinforce the experience gained in the first two labs with respect to structure visualization/analysis. By introducing a few basic CHARMM commands, this lab exercise will allow you to access the CHARMM program beyond what is directly available from the Insight II or QUANTA menus. A few basic commands from the *Atom Selection* and *Coordinate Manipulation* routines will be introduced.

During the lab you will visually inspect and identify some secondary structural elements of lysozyme. The crystal structure of lysozyme will then be minimized using the Steepest Descent (SD) minimization algorithm and the Adopted Basis Newton-Raphson (ABNR) minimization algorithm. Comparing the two methods will emphasize their qualitative features. Finally, a number of minimized lysozyme structures that were generated with different minimization schemes, are provided. You will analyze these structures using the Atom Selection and Coordinate Manipulation routines of CHARMM. These routines will be accessed through the visualization program.

II. BACKGROUND

The issues addressed in this lab concern non-bonded interaction models, energy minimization algorithms and structural analysis tools. A brief discussion of these topics is given below, more details can be found in the references.

Truncation schemes for non-bonded interactions

Every time the system is minimized or dynamically propagated its energy function has to be repeatedly evaluated. As computer power is always limited, the time it takes to evaluate the energy function often determines how long the simulation can be run. Thus, improving the efficiency of energy evaluation directly affects the possible scope of the simulations.

On closer inspection, one sees that while the bonded terms scale like N, the amount of computation that goes into evaluating the non-bonded energy terms scales like N^2, where N is the size of the system (since every atom interacts with every other atoms in the system). As a result the non-bonded interactions take the majority of the time required for energy evaluation. It is highly desirable to reduce this effort as much as possible. This is often accomplished by truncating the non-bonded interactions at a certain cutoff distance, assuming that the overall contribution of far away atoms is small. This is clearly an approximation, and one should always try to take as large a cutoff distance as possible and if possible even avoid it altogether (e.g., in small systems). Today a common cutoff distance is about 15 Å.

CUTOFF

Note, however, that a simple application of an energy cutoff will create a discontinuity in the energy function at the cutoff point. This must be avoided, since the energy discontinuities ΔV_{cut} at the cutoff point, no matter how small, introduces very large artificial forces into the system. Recall that forces are the derivatives of the energy function. Thus the sudden energy jump ΔV_{cut} introduces an infinitely large artificial force at the cutoff point. To avoid this artificial and damaging effect, energy truncation must be introduced gradually over a range of distances. In this lab you will meet two truncation schemes: Switch and Shift.

SWITCH

(i) SWITCH - In the Switch scheme the non-bonded energy terms are multiplied by the function $sw(r, r_{on}, r_{off})$, where r is the inter-atomic distance and r_{on} and r_{off} define the switching range. The functional form, which multiplies the actual energy term, is

$$sw\,(r,\,r_{on},\,r_{off}) = \begin{cases} 1 & r < r_{on} \\[2ex] \dfrac{(r_{off} - r)^2\,(r_{off} + 2r - 3r_{on})}{(r_{off} - r_{on})^2} & r_{on} < r < r_{off} \\[2ex] 0 & r > r_{off} \end{cases}$$

and its effect is shown in Figure 1. Because the switching function is limited to a relatively small range (between r_{on} and r_{off}) it has a relatively small effect on the energy value, but has a much larger effect on the forces in the switching region (recall, the force is the slope of the potential). This truncation scheme is usually applied to van der Waals interactions (sometimes also to electrostatic interactions).

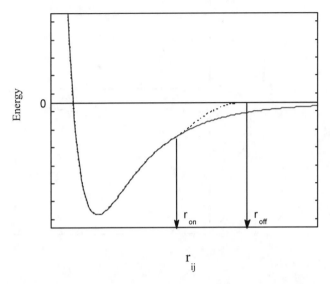

Figure 1:   The Switch truncation function applied to a van der Waals potential.

SHIFT

(ii) SHIFT - Unlike the Switch truncation, which is applied over a small range of distances, the Shift scheme affects the whole range of distances. Basically, the potential is gradually shifted towards higher energies by adding an increasing value which reaches $-\Delta V_{cut}$ at the cutoff distance r_{cut}. Figure 2 shows the effect of the Shift function on an electrostatic potential. The functional form of the Shifting function is

$$sh\,(r,\,r_{cut}) = \begin{cases} \left[1 - \dfrac{2r^2}{r_{cut}^{\,2}} + \dfrac{r^4}{r_{cut}^{\,4}} \right] & r < r_{cut} \\[2ex] 0 & r > r_{cut} \end{cases}$$

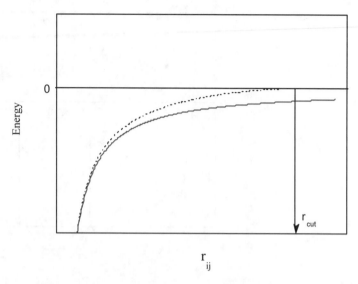

The qualitative difference between Shift and Switch is that with Shift
the forces are less perturbed although the overall accuracy of the energy
values is smaller. Because long-range forces are important with the
electrostatic potential, Shift is often applied to this interaction.

Dielectric Model
A major problem with macromolecular simulations is how to account for
the solvent, which surrounds the molecule and affects its dynamics and
thermodynamics. One way is to include an explicit model of all the water
molecules in the vicinity of the studied molecule. In this case the
electrostatic interactions between the solute and the solvent are given
CDIElectric   explicitly and a fixed dielectric constant $\varepsilon = \varepsilon_0 = 1$ can be used (CDIE =
Constant DIElectric). A constant dielectric factor ($\varepsilon = 1$) is also used in
vacuum simulations, where the solvent is absent altogether. (Recall, $E_{el} =$
$\Sigma_{ij}\, q_i\, q_j\, /\, 4\, \pi \varepsilon\, r_{ij}$).

Although including explicit water molecules is physically good, it is
computationally very expensive, as the water molecules increase the size
of the simulated system by a large factor. To overcome this problem
several simplified approaches were developed. These approaches
introduce solvation effects without actually including solvent.

RDIElectric

In the simplest of these approaches the solvent's 'screening' effect is artificially introduce by a distance dependent dielectric factor (RDIE); i.e., ε is no longer a constant but a function of the distance between the charges, $\varepsilon = \varepsilon_0 r_{ij}$. When the distance r_{ij} is large, the effective dielectric constant is large and the electrostatic interaction between two non-solvent atoms decreases. At short distances the opposite is true. The linear dependence on the distance is only for numerical convenience, other functional forms are possible too. It should be noted, that despite the crude approximation this approach gives relatively good results even for complex systems. Other, more sophisticated, approaches for introducing solvation into the simulation give more reliable results. These methods are beyond the scope of this lab.

Minimization algorithms

In CHARMM there are several alternative minimization algorithms, which can be used separately or sequentially as the task at hand may require. In general, minimization algorithms are characterized as first-derivative algorithms (i.e., algorithms that use the first derivative of the potential), second derivative algorithms (i.e., algorithms that use the second-derivative of the potential) and methods in between. In principle, the second-derivative methods are more accurate than the first-derivative methods but slower. A representative of the first derivative methods is the non-converging Steepest-Descent (SD) algorithm. A second-derivative method is Newton-Raphson (NR), and intermediate methods are represented by the Conjugate-Gradient (CG) and Adopted-Basis Newton-Raphson (ABNR) algorithms.

In this lab you will compare the non-converging first-derivative Steepest-Descent (SD) algorithm with the a converging method that makes a local harmonic approximation - Adopted-Basis Newton-Raphson (ABNR).

SD

(i) Steepest-Descent (SD)

The simplest minimization algorithm is Steepest Descent. In this procedure, at each step the gradient of the potential is calculated (i.e., the first derivative in multi-dimensions) and a displacement is added to all the coordinates in a direction opposite to the gradient (i.e., in the direction of the force). The step size is increased if a lower energy results and decreased otherwise. Because the algorithm uses finite displacement steps, it does not follow the gradient smoothly down to the minimum; i.e., it is bound to overshoot it and then jitter around it. As a consequence this method usually does not converge to the minimum, but reaches a limit cycle around it. Nevertheless, despite its poor convergence and accuracy, its 'cheapness' and its gentle positional shifts makes it useful for small changes (such as those involved in relieving bad contacts), or as the first step of a multi-step minimization scheme.

Newton-Raphson

(ii) Newton-Raphson (NR)

This is a second-derivative method, which uses the full matrix of second derivatives of the potential ($\partial^2 E/\partial r_i \partial r_j$) to solve the Newton-Raphson equations. The method adopted in CHARMM for solving these equations involves diagonalization of the second-derivative matrix and then finding the optimum step size along each eigenvector; provisions are made not to converge to saddle points. Although this minimization procedure is very accurate and converges well when starting close to a minimum it is computationally very expensive when applied to large systems.

ANBR

(iii) Adopted-Basis Newton-Raphson (ABNR)

Adopted-basis Newton-Raphson minimization (ABNR) is a generally useful first-derivative method, which is particularly suited for large systems such as proteins. Rather than using the full set of vectors as in the NR method, it adopts a smaller basis, which is limited to the subspace in which the system has made the most progress in the past moves. In this way the system moves in the best direction in a restricted subspace. For this small subspace a second-derivative matrix is constructed and solved by the NR method. Thus this method enjoys the speed and storage requirements of the first-derivative methods while retaining the most significant second-derivative information.

CONJ

(iv) Conjugated Gradients (CONJ)

Conjugated gradients minimization (CONJ) is a very useful first-derivative method, which is particularly suited for large systems such as proteins. To compensate for the missing curvature information CONJ makes use of its 'memory' of gradients calculated in previous steps. The first step is taken in the direction of the force, but for all subsequent iterations the direction of the minimization step is determined as a weighted average of the current gradient and the direction taken in the previous iteration

Gradient RMS

As the minimization process proceeds it is desirable to gauge how well it is doing and how far it is from the target minimum. Clearly the energy value itself cannot be used for this purpose as the target energy is unknown. Instead a useful measure for gauging the progress of a minimization process is the RMS value of the one-dimensional slopes, **GRMS** $\partial E/\partial r_i$ (also known as Gradient RMS [=GRMS] or RMS force). As the system nears the minimum the slope of the potential becomes smaller and is expected to reach zero at the minimum point. Thus GRMS is often used as a criterion for terminating minimization. A typical GRMS termination criterion will be in the range $10^{-3} - 10^{-6}$ kcal/mol Å (depending on the purpose of the simulation).

Geometrical characteristics

In this lab you will use two geometric measures, the Root Mean Square Distance (RMSD) and the Radius of Gyration (RGYR), to compare the structures.

The RMS Distance between two structures is an rms average over all atoms of the distance between its position in two conformations of a given molecule, after optimally superpositioning the two conformations. If $r_i^{(j)}$ is the coordinate of the i-th atom in conformation (j) and $r_i^{(k)}$ is the coordinate of the same i-th atom in conformation (k) the distance for atom i between the conformations is defined as $r_i^{(j)} - r_i^{(k)}$. Taking a square root of the average of the squares of these distances, for all (or some) atoms, results in the RMS Distance between the conformations

RMS DISTANCE

$$RMSD^{(j-k)} = \sqrt{\frac{1}{N} \sum_{i=1}^{N} \left(\mathbf{r}_i^{(j)} - \mathbf{r}_i^{(k)}\right)^2}$$

To be more precise, one is looking for the minimal RMSD possible, which means that the two structures must first be optimally overlaid, before the RMSD is calculated. This optimal overlay can be based on all atoms overlap or on a partial overlap (e.g., just the backbone).

The Radius of Gyration is one of a series of measures for the size of a molecule. It is defined as the average distance of the atoms from the molecules center of mass $\mathbf{r}_{com}$,

RADIUS OF GYRATION

$$RGYR = \sqrt{\frac{1}{N} \sum_{i=1}^{N} \left(\mathbf{r}_i - \mathbf{r}_{com}\right)^2}$$

The smaller the Radius of Gyration the more compact the molecule.

REFERENCES

1. Brooks, B.R., Bruccoleri, R.E., Olafson, B.D., States, D.J., Swaminathan, S., Karplus, M., "CHARMM: A Program for Macromolecular Energy, Minimization, and Dynamics Calculations". *J. Comput. Chem.* (1983), **4**:187-217.
2. Brooks, C.L., III, Karplus, M. & Pettitt, B.M. *Proteins: A Theoretical Perspective of Dynamics, Structure and Thermodynamics,* Advances in Chemical Physics 71 (John Wiley & Sons: New York, 1988).
3. McCammon, J.A. & Harvey, S.C. *Dynamics of Proteins and Nucleic Acids* (Cambridge University Press: Cambridge, 1987).
4. Becker, O.M., MacKerell Jr., A.D., Roux B. and Watanabe M. (Eds.), *Computational Biochemistry and Biophysics* (Marcel Dekker: New York, 2001).

III. PROCEDURE

Create a new sub-directory *lab3* (**mkdir** *lab3*), and change (**cd**) to that directory. Copy all the files from the directory *$Lab/lab3* to your *lab3* directory (**cp** *$Lab/lab3/** **.**). You will now have five files in your directory (**ls**): one pdb file (*pdb8lyz.pdb*) and four CHARMM coordinate files (*LYZ_MIN1.CRD, LYZ_MIN2.CRD, LYZ_MIN3.CRD, LYZ_MIN4.CRD*)

A. SYSTEM INITIALIZATION

1) Start QUANTA and **Import** the Protein Data Bank (PDB) file *pdb8lyz.pdb* into QUANTA (set up symmetry = NO). Since PDB files contain just the heavy atoms of the protein, no hydrogen atoms will be displayed.

2) Before we go on and place the hydrogens using CHARMM, we must set it to the right mode. From the pull-right menu under the **CHARMm Mode** option in the **CHARMm** menu, choose **RTF**. From the same menu pick **Initialization Options** and make sure the *CHARMM.LOG* file is being written.

3) To place hydrogens and calculate the energy of the crystal structure, choose **CHARMm Energy** from the *Modeling* palette. CHARMm

will add hydrogens to the molecule. Save the result to a New Filename: *8lyz_h.msf*. After saving the file the program will go on to calculate the energy (it may take a while). Record the total energy (displayed in the upper right hand corner of the screen) and the different contributions to it (displayed in the *Textport*).

4)   This is the initial structure, which will be used in future comparisons. **Export** this coordinate set as a CHARMM CRD file, naming it *8LYZ_CRY.CRD* (thus identifying it as the crystal structure). At this point you may enter a comment. We will return to this file later. Finish by clicking OK on a blank comment line.

WARNING: CHARMM has some peculiar case-sensitivities. It will accept file names only if all the letters are in the same case. Namely, it will accept *8LYZ_CRY.CRD* or *8lyz_cry.crd*, but not *8LYZ_CRY.crd* or any other case mixture of the sort.

B. VISUAL INSPECTION OF THE CRYSTAL STRUCTURE

Lysozyme has 129 amino acids and consists of two domains separated by a cleft. The enzyme consists of 15% β-sheet, 34% α-helix contained in 4 helical regions in addition to a 3_{10}-2 helix at the terminal end. Using the tools of QUANTA, find the residue ranges for the helices. Hint: Display only the backbone atoms and then create a Sheet model. To do so go to the **Solid Models/Selection Tools** option under the **Draw** menu. Click on the **Sheet** option (in the upper right corner of the large *molecule* window). Then choose **All Atoms** and **Finish**. Alternatively, you can use the **Protein Cartoon** option from under **Solid Models** in the **Draw** menu. In the *Objects Management* window you can toggle OFF/ON the Display of the sheet model. Rotate the molecule and click on individual atoms for identification. Experiment with the different graphic objects and space filling models to see the different visualizations possible in QUANTA. (To delete the ribbon structure goto the *Objects Management* window).

Q1:   From the visual analysis of the lysozyme structure, what are the residue ranges for the helices? Also identify the four S-S bonds.

C. CHARMM MINIMIZATION

1) Under the **CHARMm** menu click on **Update Parameters**. This is where you specify the scheme for treating the non-bonded interactions. Look at the fields in the dialogue box and confirm that the following scheme is set:
   - Both van der Waals interactions and electrostatic interactions use the *switch* model for smoothing.
   - Cutoff Distance = 15.0 Å, Cutoff On (CTON) = 11.0 Å and Cutoff Off (CTOF) = 14.0 Å.
   - Update frequency = -1 (which means: update as need arises).
   - Dielectric is CDIE with $\varepsilon = 1.0$.

2) Under **Minimization Options** choose the **Steepest Descent** method. (NOTE: only 50 steps of minimization are being done in this exercise. Often, hundreds to thousands of minimization steps are performed, depending on the problem at hand.) From the *Modeling* palette choose **CHARMm Minimization** (it may take a while). Record the energy and the rms force as a function of minimization step. This information is given in the *Textport*. Recalculate the **Energy** to get the component energies. Record them.

3) Now, return to the original crystal structure and redo the energy minimization using the ABNR algorithm. To get the unminimized crystal structure simply choose **Reject Changes** from the *Modeling* palette. Calculate the CHARMM energy to confirm that it is, in fact, the same structure as the initial structure. (Alternatively, you can **Open** *8lyz_h.msf* **Replacing** the previous structure). From the **Minimization Options** choose the Adopted-Basis Newton Raphson minimizer and redo the minimization. Again Record the energy and rms force as a function of the minimization step. Recalculate the component energies of the final structure.

Q2:   i)   What are the total and component energy values for the lysozyme crystal structure? And after minimization? Comment on the differences found and what may give rise to such energy differences.

             ii)   Generate a plot of energy vs. minimization-step for both the SD and ABNR results (for clarity, start the plot with the energy after 5 steps; i.e., do not include the initial energy). Qualitatively, how do the two algorithms differ?

iii) Can you suggest a way in which one can combine the two algorithms for a more efficient minimization of the structure?

4) **Reject Changes** from the *Modeling* palette and **Close** the *.msf* file.

D. ANALYSIS OF MINIMIZED STRUCTURES

In this exercise, you will use the minimized structures provided in files *LYZ_MIN(1-4).CRD*. This exercise involves sending commands directly to CHARMM via the **Send CHARMm Command** option under the **CHARMm** menu. In this initial exercise, you will calculate the rms difference between the crystal structure and the minimized structure. The four structures provided were all minimized using 500 steps of the Steepest Descent algorithm. They differ in the electrostatic scheme employed in the minimization. The details of the schemes are:

Structure (.CRD)	Smoothing (Elec.)	Cutoff Dist (CUTNB)	Cutoff On (CTON)	Cutoff Off (CTOF)	Dielectric
LYZ_MIN1	Switch	8.0Å	5.5Å	7.5Å	CDIE $\varepsilon = 1.0$
LYZ_MIN2	Shift	8.0Å	5.5Å	7.5Å	CDIE $\varepsilon = 1.0$
LYZ_MIN3	Shift	8.0Å	5.5Å	7.5Å	RDIE $\varepsilon = 1.0$
LYZ_MIN4	Shift	8.0Å	5.5Å	7.5Å	RDIE $\varepsilon = 8.0$

* All structures were calculated with "switch" smoothing on van der Waals interactions.

1) **Import** the first CHARMM coordinate set, *LYZ_MIN1.CRD*. To the question about residue ID, answer - FIRST. Pick **Update Parameters** under the **CHARMm** menu and modify the parameters to reflect the electrostatic model that was used to generate this structure (see Table). Then calculate the energy. Answer OK to the warning box and Create New Generation of your current MSF file. Record the total energy and its components.

2) Under the **CHARMm** menu, choose **Send CHARMm Command** and type in the following lines:

> *open read formatted unit 25 name 8LYZ_CRY.CRD*   [OK]
> *read coor card comp unit 25*   [OK]
> *coor orient rms select* (*type ca .or. type c .or. type n .or. type o*)
>   *end*   [OK]
> *coor rms*   [OK]
> *coor rgyr*   [OK]
> [Blank] [OK]

NOTE: CHARMM is <u>not</u> tolerant to typo errors in the <u>first four</u> <u>letters</u> of each keyword.

Type NO in response to the "send CHARMm coordinates" question.

Then terminate CHARMM interactive session by clicking on the appropriate command under **CHARMm process** in the **CHARMm menu**.

These lines demonstrate some basic CHARMM commands. The first line opens a file (in your local directory) in a <u>read only</u> format, i.e., you cannot make changes to this file. The second line imports the contents of the file into CHARMM. The keywords specify that a formatted coordinate file (*coor card*) is being read from unit 25 and the coordinates are being placed into XYZ arrays referred to as the comparison arrays (*comp*). The coordinates of *LYZ_MIN1* are stored in the main XYZ arrays. The third line addresses the *Coordinate Manipulation* routines, a powerful set of routines used in the analysis of structure and dynamics. The command initiates a series of calculations which reorient the main coordinate set (in this case the minimized coordinate set) with respect to the structure in the comparison array (in this case the crystal structure) by minimizing the rms difference between the atoms specified in the *Atom Selection* command, in this case the backbone atoms. The fourth line uses the same *Coordinate Manipulation* routine to calculate the rms difference between <u>all</u> the atoms in the minimized and crystal structures. The final statement calculates the radius of gyration for the main structure. These and other CHARMM commands will be addressed in more detail in the next lab.

The results of these calculations are placed in a *CHARMM.LOG* file in your *lab3* sub-directory. Under the **File** menu choose the **Open System Window** command. Using an editor of your choice (see the NOTE below) look at the end of the *CHARMM.LOG* file to find the results of the CHARMM calculation. Find the keywords you just typed for calculating the rms difference (COOR RMS) and radius of gyration (COOR RGYR). <u>Record</u> the results. Type **Exit** to close the System Window.

NOTE: If you don't have a preferred editor use the simple screen editor **jot** (on Silicon Graphics workstations). It is self-explanatory. To start it simply type **jot** *CHARMM.LOG* in the System Window.

3) Repeat steps 1 and 2 for each of the minimized structure provided. Note, each time you initiate a new CHARMM calculation the *CHARMM.LOG* file will be rewritten, so put the previous calculation results in a safe place before beginning a new calculation. <u>Remember</u> to change the cutoff model to the one used in the corresponding minimization calculation.

4) Finally using the above commands, calculate the radius of gyration of the crystal structure *8LYZ_CRY.CRD*. (put the crystal structure in the main coordinate set by omitting the *comp* keyword, and skip the *coor orient* and *coor rms* lines).

III. LAB WRITE-UP

Answer the two questions that appeared during the lab (marked as **Q1** and **Q2**). Then answer the following questions:

Q3: Compile your results in a table. For each structure specify the total enery, the energy components, the RMS distance to the crystal structure and the radius of gyration.

Q4: From the analysis of minimized structures, describe any correlations observed between the energy and structure, address the effect of electrostatic energy and of the cutoff schemes.

is to illustrate the use of energy
ational searching and to demonstrate
eptides and proteins. Unlike previous
QUANTA only for visualization and
ualization program as well). Most of
M as a stand-alone program.
ne the secondary structure of the
S), extract several conformations
ynamics trajectory of this peptide,
he resulting structures.

II. BACKGROUND

The simplified conformational analysis problem you are presented with in this lab is related to a real research problem: the sequence Arg-Gly-Asp-Ser (RGDS) has been shown to be responsible for the adhesive properties of several proteins like fibronectin. See: M. Cotrait et al. *Journal of Computer-Aided Molecular Design* **6** (1992), 113-130. To design synthetic molecules, which would display the same properties and be of potential therapeutic value, it is important to identify what structural characteristic of RGDS leads to its adhesive properties. A conformational search is the first step in the structural analysis of this peptide.

There are several possible ways for searching molecular conformations. One useful way is to use a high temperature molecular dynamics trajectory, which consists of a sequence of coordinate-sets in temporal order. At high temperatures, the peptide is able to overcome potential energy barriers and explore a wider region of its potential energy surface. The configurations sampled with high temperature molecular dynamics are then minimized and new minima (i.e., new stable conformations) are identified. This type of analysis is useful, for instance, when designing a drug molecule that has to fit into an active site of an enzyme.

Normally, these are time consuming calculations, requiring many CPU hours on fast computers. For the present exercise, we will just go through the general procedure.

III. PROCEDURE

Create a new sub-directory *lab4* and change to that directory. Copy all the files from the directory *$Lab/lab4* to your *lab4* directory (**cp** *$Lab/lab4/* *.*). In the *lab4* directory you will find three files and a sub-directory named *lib,* which contains two files.

NOTE: The instructions that relate to using QUANTA as a visualization program can be easily adapted to any other visualization program.

A. SYSTEM INITIALIZATION

1)   Start QUANTA and import the CHARMM coordinate file *RGDS.CRD* into QUANTA (residue ID from the first column). This is the starting structure employed in the dynamics trajectories that will be analyzed in this lab. You will learn more about dynamics in a future lab.

Q1:   Obtain the values of the dihedral angles along the backbone (you may consider displaying only the backbone atoms, for clarity). Consult the table below and identify the type of turn used as the starting structure (the angles are defined to about ±30°).

β-TURNS

Type [a]	Dihedral Angles of Two Central Residues (degree) [b]			
	Φ_2	Ψ_2	Φ_3	Ψ_3
I	-60	-30	-90	0
I'	60	30	90	0
II	-60	120	80	0
II'	60	-120	-80	0
III	-60	-30	-60	-30
III'	60	30	60	30

a)   Only the main turn types are listed, there are additional types of β-turns.

b)   The two central residues of a tetrapeptide β-turn are the *i*+1 and *i*+2 or 2nd and 3rd residues characterized by $(\Phi, \Psi)_2$ and $(\Phi, \Psi)_3$ given above.

2)   Once you have identified the structure, it is instructive to rebuild the idealized structure and compare it to that in *RGDS.CRD.* In the *Molecular Management* window turn off the display if RGDS (under **Visible**). To build your structure, select **Sequence Builder** (under

Application/Builders), select AMINO.RTF as your database and pick the RGDS sequence (Arg-Gly-Asp-Ser), choose **Set Secondary Conformation** under the **Conformation** menu. Select the residues involved in the turn. Click **OK**. Pick the backbone conformation that you think describes the molecule best and click **Done**. Leave the sequence builder by **Returning to Molecular Modeling** and save the sequence as *myrgds.seq*. Your structure is now displayed. Turn on the RGDS display and compare the two structures. Close the *myrgds.msf* and continue.

3)   **Update Parameters** (under the **CHARMm** menu) so that the interaction energy is calculated without any cutoffs. To do so set the cutoff distance along with its On and Off parameters to a sufficiently long distance - <u>much larger</u> than the molecule's size. Set the <u>Image Cutoff Distance</u> to a value larger than the Cutoff Distance (this is required by QUANTA, even though you won't use the image facility in the lab).

4)   Calculate the energy. <u>Record</u> the total energy and the different contributions to it as displayed in the *Textport*. (in case of missing parameters just **quit** to accept the defaults suggested by QUANTA).

B. ANIMATION OF A DYNAMICS TRAJECTORY

While you will not actually calculate a molecular dynamics trajectory in this lab, you will animate and analyze such a trajectory. Recall, that a dynamics trajectory consists of a sequence coordinate sets representing the structure of the molecule at different times. By displaying the coordinate sets one after the other an animation of the motion is obtained.

DYNAMICS
TRAJECTORY

1)   Under the **Application** menu choose the **Dynamics Animation** option. A new *Dynamics Animation* palette appears on the right hand side of the screen. Choose **Select Trajectories** followed by **Initialize dynamics files.** QUANTA will ask you which trajectory you want to visualize. There are two trajectory files in the *lab4* directory. Both are vacuum simulations of the RGDS peptide performed at the high temperature of 1000 K, which differ only in the electrostatic model used. The first file, *DYNA_EPS1.DCD*, contains a trajectory generated by a simulation in which a constant dielectric of $\varepsilon = 1$ was used. The second file, *DYNA_EPS8R.DCD*, contains the result of a simulation in which a distance dependent dielectric factor was used ($\varepsilon = 8*r$, where

r is the charge-charge distance). The electrostatic interactions were not truncated (i.e. a very large cutoff was used, so that all interactions are included in the calculation). Chose the first file, *DYNA_EPS1.DCD,* and then **Exit**.

2)   This particular simulations extend for 20 picoseconds, between *t* =7 ps and *t* = 27 ps. The first seven ps of the simulation were taken up by the heating and equilibration phases of the simulations. These are not used in the analysis and are thus not included in the trajectory file. A "snapshot" (or "time frame" or "data set") was written every 10 femtosecond (10^{-14} s), so that the trajectory file contains 2000 data sets. This information is displayed in the *Textport*: you can see there that the trajectory contains 2000 data sets (i.e. 2000 conformations) starting with data set 7010 and ending at data set 27000. You will learn how to generate a dynamics trajectory in a later lab.

Click on **Set Up Animation**, set the Dataset Range from 7010 to 27000, with a step size of 10 (this will include all data sets in the animation), and set Clock Speed to 500. Now select **Create Animation** from the palette. This downloads the 2000 trajectory frames from the file onto the central memory of the computer (do not interrupt this process even if it takes a while).

3)   Once you have created the animation, you can view it again by picking the **Clock** option (which displays the trajectory from beginning to end, then jumps back to the beginning and continues) or the **Cycle** option (when it reaches the end of the trajectory, it returns to the beginning by displaying the trajectory in reverse). The small *dial* window allows you to control the speed of the animation while you view it. You can pause the animation at a particular frame and restart it again by clicking on **Clock/Cycle**. You can rotate and move the molecule during the animation.

Q2:   Give a qualitative description of what you observe in the dynamics trajectory. Note in particular: large conformational changes, which residues appear to be more "mobile", which residues appear to interact with each other, etc. You may find it useful to view the motion of the backbone atoms alone and to slow down the dynamics animation. To do so display the backbone atoms of your starting structure before you download the dynamics trajectory.

Q3:　　　　Using the facilities available in QUANTA, calculate the inter-atomic distances between the following atoms every 100 time frames for at least 5 ps. Plot these results as a function of time. Compare these values with those of the initial *RGDS.CRD*

Residue 1 atom N　　　-- Residue 4 atom C
Residue 1 atom O　　　-- Residue 4 atom H
Residue 1 atom CZ　　 -- Residue 3 atom CG

4)　　Extract selected conformations from the dynamics trajectory you just viewed. These conformations will be later subjected to a energy minimization using CHARMM as a stand alone program. From the *Dynamics Animation* palette choose **Write Frame to MSF...** . Save the data set number 7910 (t = 7.91ps) to a new msf file, named *7910_eps1.msf* and **Retain your previous selection**. Then write two more MSF files for the data sets 10330 (t = 10.33ps) and 19810 (t = 19.81ps). Exit the dynamics animation and by opening the newly generated msf files, export <u>each of them</u> as CHARMM coordinate files; e.g., *7910_eps1.crd*. You will use these in the CHARMM minimization.

WARNING: CHARMM has a peculiar case sensitivity. File names must be in a single case (all upper case or all lower case). Mixed case file names will not be recognized (e.g., *7910_eps1.CRD* will not work).

5)　　Repeat the whole exercise for the second trajectory *DYNA_EPS8R.DCD* (including **Q1** and **Q2**; can you relate the difference between the trajectories to the electrostatic models used?)

OPTIONAL: Save the data sets 9320 (9.32 ps); 12160 (12.16 ps); 15190 (15.19 ps) and 21990 (21.99 ps) for future use.

C. PREPARATION of CHARMM INPUT FILES

> Read the **CHARMM Primer** section at the beginning
> of this *Guide* before continuing.
> **It is essential!**

Now you will work with the stand-alone CHARMM program to minimize
these structures and analyze them (calculate the values of the dihedral
angles along the backbone). For an initial introduction to CHARMM and
how to generate your own CHARMM input file (*charmm.inp*) refer to the
CHARMM Primer provided in this *Guide*.

This exercise has two parts:

1)   In the first part of this exercise you will prepare the CHARMM
      command file according to the detailed instructions given below.
      This command file will perform energy and structural analysis of
      the original RGDS structure.

2)   In the second part you will modify the command file you just
      created. This time the analysis will be performed on the
      conformations that were extracted from the molecular dynamics
      trajectory. Unlike the first part, these conformations are going to be
      minimized *before* performing the analysis.

1) Analysis of initial RGDS structure

Consult the **CHARMM Primer** and follow the detailed instructions below
to generate the command file that will later run as a batch job. The
complete file is given at the end of the **CHARMM Primer**.

     Start by opening a new file named *charmm.inp* (use whatever editor
you like; **vi**, **emacs** or **jot**).

a)   To give this job a title type on the first lines of the file something like
      this:

      ```
      * First Biophysics job
      *
      ```

When executed the program will answer with RDTITL> and echo each line you type in. In a batch mode this printout is directed to the output file; in an interactive mode it is directed to the screen).

b) Now type the following lines to read the RTF and PARAmeter files:

RTF

```
OPEN READ CARD UNIT 25 NAME lib/toph19.inp
READ RTF  CARD UNIT 25
OPEN READ CARD UNIT 26 NAME lib/param19.inp
```
PARAmeters
```
READ PARAmeter CARD UNIT 26
```

Note the similarity in the format of these two statements and the statements used in the previous lab to read in the coordinate files.

COMMENT: The UNIT number is an arbitrary internal identification number of the opened file (borrowed from FORTRAN).

c) Generate a PSF for the *RGDS* peptide. Use the SETUP keyword to setup an IC Table:

```
READ SEQUence CARD
* RGDS
*
```
GENErate
```
4
ARG GLY ASP SER
GENErate RGDS SETUp
```

NOTE: CHARMM automatically puts (PATChes) an amino and a carboxylate groups on either ends of the polypeptide chain.

d) We can now read in the original coordinate set:

COORdinate
```
OPEN READ FORMatted UNIT 27 NAME RGDS.CRD
READ COORdinate CARD UNIT 27
```

COMMENT: FORMatted and CARD both mean that the file to be read is a "text" file and not a "binary" file.

e) Now use the NBONd command to specify how the non-bonded interactions are to be treated:

NBONd
```
NBONd CUTNb 150.0 CTONnb 145.0 CTOFnb 149.0 SHIFt -
    VSWItch CDIElectric EPSIlon 1.0
```

The CTONnb and CTOFnb that appear in the NBONd command correspond to the CTON and CTOF used in QUANTA, respectively. SHIFt = "shift" truncation scheme applied to the electrostatic interactions; VSWItch = "switch" truncation scheme applied to the van der Waals interactions; CDIElectric EPSilon 1.0 = use a constant dielectric factor, $\varepsilon = 1$. The <u>hyphen</u> at the end of the first line indicates that the command continues on the next line.

f)   Now to calculate the energy type,

ENERgy

```
ENERgy
```

g)   Finally, based on the coordinate you read in, fill the IC Table and print it (to the output file):

IC

```
IC FILL
PRINT IC
```

i)   To terminate a CHARMM job type:

```
STOP
```

Follow the instruction in the ***CHARMM Primer*** and execute this job in batch mode, by typing at the UNIX prompt:

charmm < *charmm.inp* > *charmm.out*

To read the content of the output file use either an editor (**vi**, **emacs** or **jot**) or display it on the screen (use the UNIX commands **more** or **cat**).

2)   Analysis of the RGDS trajectory

In this part of the exercise you will use CHARMM to perform a similar analysis on the three conformations you extracted from the *DYNA_EPS1.DCD* trajectory file. The only difference being that in this analysis you will minimize the structures (to their nearest local minimum) before performing the analysis. This minimization removes thermal fluctuations and highlights of the underlying stable conformations.

Since the procedure is very similar to the one used in the previous section you are advised to modify the command file you just created. To do so follow the instructions below:

a)   Copy your CHARMM command file to a new file; e.g., *analyze.inp* (**cp** *charmm.inp analyze.inp*). Use an editor of your choice to open the new file for editing.

b)   On the first line of the file change the title of the job to describe the current task.

c)   Instead of reading the coordinates of the original structure (*RGDS.CRD*) read in one of the CHARMM coordinate files that you extracted from the trajectory (e.g., *7910_eps1.crd*).

d)   After the ENERgy command (which will now give you the energy of the unminimized structure) and before filling the IC Table, add the following two lines. These instructs the program to perform 500 steps of Steepest Descent minimization (printing out every 50 steps) and then to get the energy of the minimized structure:

MINImize
SD

```
MINImize SD NSTEp 500 NPRInt 50
ENERgy
```

After making these modifications execute the new command file using CHARMM in a batch mode. Be sure to direct your output to a different output file.

Perform the same calculation on the two other structures you extracted from the trajectory. Each time change the name of the coordinate file in the input file and direct the results to a different output file.

OPTIONAL: perform the same analysis on the structures extracted from the *DYNA_EPS8R.DCD* trajectory.

IV. LAB WRITE-UP

Answer questions **Q1** - **Q3**, which were posed earlier in this lab. In addition answer the following questions:

Q4   What are the values of the dihedral angles (ϕ, ψ, χ) along the backbone for the three structures you minimized (*7910_eps1.crd*, *10330_eps1.crd* and *19810_eps1.crd*)? Mark the backbone (ϕ, ψ) angles for residues 2 and 3 as points on a (ϕ, ψ) plane, and compare these values to a standard Ramanchandran plot. In what structural regions do these simulated structures fall. Mark the original structure on this plot too.

Q5   Compare the energy of the three minimized structures and the original *RGDS.CRD* (both total energy and its different contributions). Comment on the results.

Q6   OPTIONAL: As you have seen, the conformations obtained from a simulation depend on the electrostatic model used. From the calculated energies of the minimized conformations 7910_eps1; 9320_eps8r; 12160_eps8r, what are the dominant contributions to the energies? Of the two simulations, which do you think is the more realistic, i.e. which one gave more reasonable structures, which one better described the interactions in a protein/water environment? Why? [Refer to M .Cotrait et al. *Journal of Computer-Aided Molecular Design* (1992), **6**:113-130 for discussion on this point].

Lab 5:   Basic Molecular Dynamics
                  in Vacuum and in Solution

I. OBJECTIVE

This lab is the first in a series of labs dealing with molecular dynamics of biomolecules. While in the previous lab you used prepared dynamic trajectories of a small peptide, in this lab you will learn how to generate such a trajectories, both in vacuum and in solution.

The main goal of this lab is to highlight the different steps involved in setting up a molecular dynamics simulation: generation of the model, preparation, equilibration, production and analysis. For this purpose this lab focuses on simulations of small molecular systems. In later labs you will perform dynamic simulations of larger and more complex molecular systems.

II. BACKGROUND

Molecular Dynamics

Molecular dynamics simulations calculate the temporal evolution of a molecular system by numerically integrating the system's classical equation of motion (Newton's equation, $F = ma$). Namely, the following N coupled differential equations

NEWTON'S EQUATION OF MOTION

$$m_i\frac{d^2\mathbf{r}_i}{dt^2} = \mathrm{F}_i = -\nabla_i V(\mathbf{r}_1,\mathbf{r}_2,...,\mathbf{r}_N) \qquad i = 1, N$$

are simultaneously solved to obtain the atomic positions and velocities as a function of time. Here m_i and $\mathbf{r}_i$ are the mass and position of particle i, and $V(\mathbf{r}_1, \mathbf{r}_2, ..., \mathbf{r}_N)$ is the potential energy function, that depends on the positions of all the N particles in the system. The potential energy function was discussed in length in Lab 2. The negative gradient $-\nabla_i V(...)$ of the potential with respect to the position of particle i is the force F_i acting on that particle.

To solve this set of coupled differential equations it is necessary to define the initial conditions from which the integration starts. These initial conditions include the initial positions of all atoms and their velocities at the initial time. For protein simulations the initial positions are usually based on known X-ray or NMR structures (followed by energy minimization to relieve local strain due to non-bonded overlaps and distortions). The initial velocities are typically assigned from a Maxwellian distribution at some relatively low temperature, namely from **MAXWELL-** a Gaussian distribution in each of the three velocity components v_x, v_y and **BOLTZMANN** v_x corresponding to a Maxwell-Boltzmann distribution in velocity space. The system is then slowly heated up to the desired simulation temperature. The temperature at any given moment is defined in terms of the mean kinetic energy

TEMPERATURE

$$T(t) = \frac{1}{k_B M} \sum_{i=1}^{M} m_i |v_i|^2$$

where M is the total number of unconstrained degrees of freedom, v_i is the velocity of particle i at time t, and k_B is the Boltzmann constant.

Every molecular dynamics simulation is divided into several stages:

MOLECULAR DYNAMICS PROCEDURE

i)   Model Generation: The first step involves generating the molecular model and setting-up the initial conditions. In terms of CHARMM it means first generating the molecular PSF (*Protein Structure File*) and then reading in the initial set of coordinates, usually based on X-ray crystallography or NMR spectroscopy.

ii) Preparation: In this step the system is brought to a state where actual dynamics can be performed. This typically involves minimization of the initial structure to relieve local strain and then gradually heating-up the system to the desired temperature.

iii) Equilibration: This is an essential stage in which long dynamic trajectories are run to make sure that the system is equilibrated. In most studies (though not all) this is a prerequisite for reliable results.

iv) Production run: The actual dynamics simulation from which data is accumulated.

v)   Analysis: An essential step, which converts the raw data accumulated in the production run to meaningful information!

Integration Algorithms

A good starting point for understanding finite difference methods is the Taylor expansion about time t of the position at time $t + \Delta t$,

INTEGRATORS
$$r(t + \Delta t) = r(t) + \dot{r}(t)\Delta t + \tfrac{1}{2}\ddot{r}(t)\Delta t^2 + \ldots$$

alternatively this can be written as

$$r(t + \Delta t) = r(t) + v(t)\Delta t + \tfrac{1}{2}a(t)\Delta t^2 + \ldots$$

where $v(t)$ is velocity vector and $a(t)$ is the acceleration. Since the integration proceeds in a stepwise fashion, and recalling equation (3.6), it is convenient to rewrite the above expansion in a discrete form. Using r_n to indicate the position at step number n (at time t) and r_{n+1} to indicate the position at the next step n+1 (at time $t + \Delta t$) equation (3.12) can be written as

$$r_{n+1} = r_n + v_n \Delta t + (F_n / 2m)\Delta t^2 + O(\Delta t^3) \qquad (3.13)$$

with this information the velocity v_{n+1} at time n+1 can be crudely estimated, for example, as

$$v_{n+1} = (r_{n+1} - r_n)/2 \qquad (3.14)$$

Together, equations (3.13) and (3.14) form an integration *algorithm*. Given the position r_n, the velocity v_n and the force F_n at step n, these equations allow one to calculate (actually, estimate) the position r_{n+1} and velocity v_{n+1} at step $n+1$. The formulation is highly trivial and results in a low quality integration algorithm (large errors). Other, more accurate, algorithms have been developed using the same kind of reasoning. In the next sub-sections we shall survey some of the more commonly used finite difference integration algorithms, highlighting their advantages and disadvantages.

Verlet Integrator

VERLET

The most common integration algorithm used in the study of biomolecules is due to Verlet. The Verlet integrator is based on two Taylor expansions, a forward expansion $(t + \Delta t)$ and a backward expansion $(t - \Delta t)$.

At the heart of all molecular dynamics simulations are the algorithms for numerically integrating of the equations of motion. These integration algorithms, which typically propagate the motions step by step, differ from one another in accuracy and speed. For the purpose of macromolecular dynamics one usually uses second order "finite difference" methods, such as the *Verlet* or the *leap-frog* integrators, although at times higher order integrators may also be used. Second order integration algorithms are based on a Taylor expansion of the variables. For example, in the *Verlet* integrator, which is the simplest finite difference integrator, the position at time $(t + \Delta t)$ is expanded in terms of the information at time t, i.e.,

$$r(t + \Delta t) = r(t) + \dot{r}(t)\,\Delta t + \frac{1}{2}\ddot{r}(t)\,\Delta t^2 + \dots$$

As the propagation is done using discrete time steps Δt, the time t is associated with step n and time $(t + \Delta t)$ is associated with step $n+1$. Thus the above expansion can be rewritten as

$$r_{n+1} = r_n + v_n\,\Delta t + \frac{F_n}{2\,m}\,\Delta t^2 + O(\Delta t^3)$$

where the first derivative of the position is the velocity v_n, and the second derivative of the position is the acceleration a_n, here given in terms of force over mass ($a_n = F_n/m$).

Now, consider two such expansions, one for step $n-1$ and one for step $n+1$,

$$r_{n+1} = r_n + v_n\,\Delta t + \frac{F_n}{2\,m}\,\Delta t^2 + \dots$$

$$r_{n-1} = r_n - v_n\,\Delta t + \frac{F_n}{2\,m}\,\Delta t^2 - \dots$$

Adding and subtracting these two equations yields the basic equations of the *Verlet* integrator,

$$r_{n+1} = 2\,r_n - r_{n-1} + \frac{F_n}{m}\,\Delta t^2 + O(\Delta t^4)$$

$$v_n = \frac{(r_{n+1} - r_{n-1})}{2\,\Delta t} + O(\Delta t^3)$$

This means that the position at the next step, $n+1$, is calculated based on the position at the two previous steps (n and $n-1$) and on the force at step n. After the position is calculated the velocity is obtained based on it.

This is only one basic (though commonly used) integration algorithm. There are many other algorithms, some significantly better than the simple *Verlet* integrator, and there is much room for discussion on pros and cons of the different methods. Such a discussion, however, is beyond the scope of this introduction. Additional information and a detailed discussion can be found in Allen and Tildsley (1989).

Quality of the Integration

CONVER-GENCE

When performing numerical integration it is important to be able to assess the quality of the procedure (its stability). Since Newton's equations conserve energy (microcanonic system) a good simulation is expected to maintain the total energy fixed. One measure for this stability is the ratio between the average fluctuation of the total energy $<\Delta E> = <E(t) -<E>>$ and the average total energy $<E>$ of the system; i.e., $<\Delta E>/<E>$. If this ratio is smaller than 10^{-3} the simulation is considered stable. It is customary to take the decimal logarithm of this ratio, $\text{Log}_{10} (<\Delta E>/<E>)$. If this Log is smaller than -3 the simulation is stable.

Solvent Effects

An important part of setting up a simulation is deciding what environment is the molecule of interest to be simulated in. The simplest option, a vacuum simulation, is easy to perform but is often not sufficiently accurate as it neglects environmental effects. In previous labs (Lab 3 and 4) we introduced one way of improving over the vacuum results by using a distance dependent dielectric constant, which qualitatively accounts for solvent 'screening'. In this lab we go a step further and add explicit solvent molecules (in our case water) around our molecule of interest.

TIP3P WATER

Of the several optional water models we use a straightforward model, called TIP3P. In this model the three water atoms are represented as van der Waals spheres with point charges at their centers. This model is known to reproduce the properties of real water. The water 'molecules' are now added to the simulation by overlaying a box or a sphere of pre-equilibrated water on top of the molecule of interest.

BOUNDARY CONDITIONS

An important issue, which arises when dealing with explicit solvent models is that of the boundary conditions. There must be something in the simulation to 'hold' the water molecules together or they will fly apart and fail to represent a liquid. To solve this problem boundary conditions are usually imposed. The most common boundary conditions are the 'periodic' boundary condition and the 'spherical' boundary condition. A 'periodic' boundary condition means that there is a central 'box' of solvent and solute molecules, which is surrounded by periodic self-images in all directions.

PERIODIC BOUNDARY CONDITIONS

The simulation is performed on the molecules in the central cell but the forces acting on molecules in this central cell take into account the periodic images mimicking an much larger system. The resulting trajectory essentially corresponds to an *infinite* periodic system. Another option is to use 'spherical' boundary conditions, in which a force is applied on the boundary of the water sphere. This force represents the (average) interaction of boundary water molecules with a bulk of water, and thus mimics the effect of bulk water beyond the explicit water sphere.

Some selected REFERENCES:

1.   Brooks, B.R. et al. *J. Comp. Chem.* (1983), **4**: 187–217.
2.   Brooks III, C.L., Karplus M. and Pettitt, B.M. *Proteins: A Theoretical Perspective of Dynamics, Structure and Thermodynamics,* Advances in Chemical Physics 71 (John Wiley, New York, 1988).
3.   Brooks III, C.L., Karplus, M., "Solvent effects on protein motion and protein effects on solvent motion: Dynamics of the active site region of lysozyme", *J. Mol. Biol.* (1989), **208**: 159–181.
4.   Pomès R. and McCammon, J.A. "Mass and step length optimization for the calculation of equilibrium properties by molecular dynamics simulation", *Chem. Phys. Lett.* (1990), **166**: 425–428.
5.   Allen, M.P. and Tildsley, D.J. *Computer Simulations of Liquids*, 2nd edition (Oxford University Press: Oxford, 1989).
6.   Becker, O.M., MacKerell Jr., A.D., Roux B. and Watanabe M. (Eds.), *Computational Biochemistry and Biophysics* (Marcel Dekker: New York, 2001).

III. PROCEDURE

Create a new sub-directory *lab5* and change to that directory. Copy the content of the *$Lab/lab5* directory to the new directory you just created (**cp -r** *$Lab/lab5/** .). Type **ls** to see the content of this directory, which includes eight files plus a *lib* sub directory. The content of the *MEOH* files is printed out at the end of the Lab 5 chapter.

   Note, in the first part of this lab you are asked to follow detailed instructions and write the CHARMM command files yourself. This will give you a flavor of all the steps involved in the simulation. In future labs you will be given *template* files that only require slight modifications.

A. VACUUM SIMULATION

As discussed above, every molecular dynamics simulation is subdivided to several steps. In this part of the lab you will create command files for performing the different steps of the simulation for a RGDS peptide in vacuum.

1) Model Generation, Preparation and Equilibration

The first steps in every molecular dynamics study is to generate the model, load the initial coordinates, minimize them, gradually heat the system up to 300 K and then equilibrate the system. You will perform all these steps in a single job.

Start by opening a new file named *RGDS_init.inp*, use whatever editor you like (**jot**, **vi** or **emacs**). This file will hold the appropriate CHARMM commands after you type them in.

NOTE: You can use the CHARMM input file you wrote in Lab 4 (*charmm.inp*) as a template for this file.

Before you begin read
the ***CHARMM Primer*** again

More details can be found in the CHARMM documentation.

1.   Give the job a title and load the topology (RTF) and parameter (PARAM) files:

```
* Lab 5: preparation of vacuum RGDS
*
OPEN READ CARD UNIT 1 NAME lib/toph19.inp
READ RTF   CARD UNIT 1
OPEN READ CARD UNIT 2 NAME lib/param19.inp
READ PARAmeter CARD UNIT 2
```

2.   Generate a PSF for RGDS and save it to the file *rgds.psf*. Then read in the initial coordinates.

NOTE: CHARMM automatically attaches a carboxylate group and an amino group to the peptide terminals.

```
! generate PSF
READ SEQUence CARD
* RGDS
*
4
ARG GLY ASP SER
GENErate RGDS SETUp
```

```
! write PSF
OPEN WRITe CARD UNIT 3 NAME rgds.psf
WRITe PSF CARD UNIT 3

! read coordinates
OPEN READ CARD UNIT 4 NAME RGDS.CRD
READ COORdinate CARD UNIT 4
```

3. Setup the non-bonded interaction model, then calculate energy, minimize (using 500 steps of steepest descent followed by 500 steps of ABNR, printing out every 50 steps), and recalculate the energy (see details in Lab 4):

```
! non bonded interaction
NBONd CUTNb 150.0 CTONnb 145.0 CTOFnb 149.0 SHIFt -
     VSWItch CDIElectric EPSIlon 1.0
```

ENERgy
MINImize
SD
ABNR
```
! minimize initial structure
ENERgy
MINImize SD   NSTEp 500 NPRInt 50
MINImize ABNR NSTEp 500 NPRInt 50
ENERgy
```

4. Gradually heat up the system to 300 K. Start at 0 K and heat up the molecule to 300 K using 30 K increments every 500 fs (= 0.5 ps). The whole heating up process takes 5 ps.

This process is performed using CHARMM's *DYNAmics* command with the VERLET integration algorithm (STRT stands for "starting" the trajectory):

```
! open the restart file
OPEN WRITe FORMatted UNIT 10 NAME rgds.res
```

DYNAmics
```
! heat the system from 0 K to 300 K
DYNAmics     VERLet STRT -
    NSTEp  5000   TIMEstep 0.001 -
    FIRSTT  0.0 FINALT   300.0   TEMInc 30.0 -
    IHTFrq  500 NTRFrq    1000   IHBFrq   -1 -
    NPRInt   50 IPRFrq     200 -
    IASORS    1 IASVEL       1 -
    IUNWrite 10
```

The different keywords specify how the dynamics (heating up in this case) is executed:

FIRSt
FINAlt
TEMInc
- In this case we start at 0 K and heat up to 300 K using 30 K increments (FIRSTT, FINALT, TEMInc).

NSTEp
TIMEstep
- The simulation is run for 2000 steps using a 0.001 picosecond time steps (NSTEp, TIMEstep).

IHTFrq
- IHTfrq indicates that a temperature increment will be performed every 200 integration steps.

- Since the velocities are randomly chosen (from a Gaussian distribution) there is no built in mechanism to prevent the molecule to rotate as a rigid object and translate in the external reference frame. These motions are meaningless as we want to study the relative motions <u>within</u> the molecule. The keyword

NTRFrq
NTRFrq instructs CHARMM to "stop" this global rotation-translation every 1000 steps. After a few such "treatments" only internal motions will remain.

NPRInt
IPRFrq
- The energies are written out every 50 steps (NPRInt) and their averages are calculated every 200 steps (IPRfrq).

IUNWrite
- IUNWrite indicates that the information, needed to "restart" the trajectory in the next stage exactly where it finished here, will be written to the file indicated as "unit" number 10. This file was opened as *rgds.res*.

- The other four (somewhat obscure) keywords describe features such as the initial distribution of velocities.

NOTE: The hyphens (-) at the end of the lines are continuation markers and are essential for a successful execution of the job. Do not omit them.

5.   Equilibrate the system for another 5 ps at 300 K. We are continuing the same trajectory (RESTarting from the end of the previous trajectory), only this time there is no additional heating up of the system.

```
! reopen restart files
OPEN READ  FORMatted UNIT 10 NAME rgds.res
OPEN WRITe FORMatted UNIT 11 NAME rgds.res1

! equilibrate the molecule for 5 ps
DYNAmics   VERLet RESTart -
           NSTEp    5000  TIMEstep 0.001 -
           NPRInt     50  IPRFrq     200 -
```

DYNAmics

```
IHTFrq        0  -
IUNRead      10  IUNWrite       11
```

In this part of the job you are reading the "restart" file you saved at the end of the heating part (*rgds.res*) by re-OPENing it as a READ only file (UNIT 10; keyword IUNRead), you will write a new "restart" file (rgds.res1) to save the configuration at the end of this equilibration part (UNIT 11; keyword IUNWrite).

6.   Write out the coordinates at the end of the preparation phase (*rgds_1.crd*).

```
! write coordinates
OPEN WRITe CARD UNIT 15 NAME rgds_1.crd
WRITe COORdinate CARD UNIT 15

STOP
```

2)   Executing and analyzing the 'preparation phase'

Now that you have created the CHARMM command file to perform the preparation phase, execute it in a batch mode.

To execute the job type the following at the UNIX prompt,

charmm < *RGDS_init.inp* > *RGDS_init.out* &

It may take up to a few minutes to run, depending on the computer. The output will be written to this *RGDS_init.out* file.

When the job is done perform the following analysis on the results:

i)   Look in the output file (use the command **more** to scroll it screen by screen; i.e., **more** *RGDS_init.out*). Locate the first energy evaluation, the minimization energies and the second energy minimization.

Q1:   Write down the energy before and after the minimization. Recall from Lab 4 that the coordinates in *RGDS.CRD* were generated as an ideal β-turn. Based on your results, what can you say about the stability of this ideal β-turn structure?

ii) Extract the average energies that were calculated every 200 steps (0.2 ps), use the **grep** command to do so. Direct the results to a file for future use.

grep 'AVER>' *RGDS_init.out > grep_output*

Look in this file. The columns contain the following information

AVER> Steps Time E_total E_kin E_pot Temperature

Q2:   Plot the graphs of the Total Energy, Kinetic energy, Potential energy and Temperature as a function of time (Recall that in the first 5 ps you heated up the system while in the last 5 ps you just let the system equilibrate). Describe and explain the different behavior of the total energy in the two phases of the simulation.

Q3:   OPTIONAL: If you can read the file generated after the **grep** command directly into a plotting program repeat the previous **grep** only this time with the keyword 'DYNA>' instead of 'AVER>' (direct it to a different file name). This will extract the instantaneous energy values every 50 fs. Plot the graphs of the Total Energy, Kinetic energy, Potential energy and Temperature as a function of time. Compare these results to the plots of the average values (in **Q2**). Comment of the fluctuations. As you can see, the total energy fluctuates much less than the other quantities, what is the reason for that?

iii) Use QUANTA to compare the initial conformation (*RGDS.CRD*), which is an ideal β-turn, and the final conformation (*rgds_1.crd*). (Of course, you may use other visualization programs instead of QUANTA).

- Start QUANTA by typing **quanta**.
- **Import** the two CHARMM coordinate sets (one at a time).
- From the **Draw/Color Atoms** menu choose the color **by molecule** option.
- Make the original coordinate set (*RGDS.CRD*) not active and not visible (in the *Molecular Management* window).

- Use the **Hydrogen Bonds** option from the *Modeling* pallet, and click on the atoms involved in H-bonds to identify the residues involved in these bonds (detailed information in the *Textport*).

Q4:   What are the main differences between the two structures?

- **Close** all open MSF files.
- At this point you can either exit QUANTA or 'iconize' it by clicking on the box at the upper right corner of the *molecule* window. It will be some time before QUANTA will be used again later in this lab.

3) Analyze the quality of the simulation

The quality of a molecular dynamics simulation is measured by the fluctuations of the total energy. Since solving Newton equations conserves the total energy of the system, a good numeric integration should have very small fluctuations. It is customary to say that the ratio between the average fluctuation of the total energy and the average total energy should be less than 10^{-3}, or as it is often presented the $\log_{10}$ of this ratio should be smaller than -3.

These quantities can be calculated using available tools (e.g., Excel) according to the following formula:

Average of $E = <E>$

Average fluctuation of $E = <\Delta E> = (1/n)\Sigma_n(E_i - <E>)$

$\log\Delta E = \log_{10}(|<\Delta E>/<E>|)$

Apply this calculation to the output of the **grep** 'AVER>' command. Use only the last 4 ps of the simulation (starting at time 6.0 ps).

Q5:   i)   Report the results of the **awk** script.
          ii)   What is the quality of the simulation based of the fluctuations of the total energy? Compare it to the fluctuations of the other three quantities. How significant are these fluctuations in term of the quality of the simulation?

4) Production run

Now that the system is properly prepared lets make a production run. To do this type the following CHARMM script file as *RGDS_run.inp* (you can use your last CHARMM script file as a template for this one):

```
* Lab 5: Production run for RGDS
*

! read topology and parameter files
OPEN READ CARD UNIT 1 NAME lib/toph19.inp
READ RTF  CARD UNIT 1
OPEN READ CARD UNIT 2 NAME lib/param19.inp
READ PARAmeter CARD UNIT 2

! Load PSF
OPEN READ CARD UNIT 3 NAME rgds.psf
READ PSF  CARD UNIT 3

! read coordinates
OPEN READ CARD UNIT 4 NAME rgds_1.crd
READ COORdinate CARD UNIT 4

! non bonded interaction
NBONd CUTNb 150.0 CTONnb 145.0 CTOFnb 149.0 SHIFt -
      VSWItch CDIElectric EPSilon 1.0

! open restart files
OPEN READ  FORMatted UNIT 10 NAME rgds.res1
OPEN WRITe FORMatted UNIT 11 NAME rgds.res2

! open a trajectory file
OPEN WRITe UNFORMatted UNIT 12 NAME rgds.trj

! run production dynamics for 10 ps
DYNAamics VERLet RESTart -
         NSTEp    10000      TIMEstep  0.001 -
         NPRInt     100      IPRFrq      250 -
         NSAVC      250      IUNCRD       12 -
         IUNRead     10      IUNWrite     11

STOP
```

NOTE: 1) Double check that you typed the correct file names.

2) In this job you are saving a dynamic trajectory to the file *rgds.trj*. This file is opened as "unit" 12, and will include a series of conformations taken every 250 integration

steps, which is equivalent to 0.25 ps (the keywords
NSAVC and IUNCRD in the DYNAmics command).

3)   The energies are printed every 100 steps and their
averages every 250 steps (NPRInt and IPRFrq).

Run the RGDS_run.inp script file as described above,

charmm < *RGDS_run.inp* > *RGDS_run.out &*

when the job is done analyze its quality as before; i.e., **grep** for the lines
with 'AVER>' and perform the same analysis as before to get the Log of
the energy fluctuations. This time we shall asses the fluctuations over the
while 10 ps of simulations (starting at time 10 ps and ending at time 20 ps,
recall this is a continuation of the previous trajectory) and not only over the
last 4 ps as in the preparation phase.

grep 'AVER>' *RGDS_run.out* > *<grep_output>*

Q6:   Answer **Q5** again for the new results.

5) Basic analysis of the production run

The final goal of any simulation is in the analysis of its results. There are
many ways to analyze a dynamic trajectory, and you will leans several
different methods in the following labs. Here we will limit ourselves to the
most straightforward analysis technique - the Time Series. Namely,
plotting the value of a given property as a function of time and calculating
its average and fluctuations. In fact, earlier in this lab you have already
used this methodology when you plotted time series for the different
Energies of the system.

Analyzing a molecular dynamics trajectories we are often interested
in the temporal behavior of different geometric properties such as bond
lengths, angle values and torsions.

In the *lab5* directory you can find a CHARMM input file
(*RGDS_anal.inp*) that has been prepared for you. This command file
calculates the time series of the four central dihedral angles of RGDS (ϕ_2

and ψ2 around the Ca of residue 2 (Gly) and φ3, ψ3 around the Cα of residue 3 (Asp)).

Look in the prepared input file to see what it is doing. Execute this CHARMM job as in the above,

charmm *< RGDS_anal.inp > RGDS_anal.out*

The job will produce a new file named *rgds.anal* which contains five columns as follows:

```
Time   Phi2   Psi2   Phi3   Psi3
```

Q7:   Look in the job's output file (*RGDS_anal.out*) and extract the information about the average value and rms fluctuations of each of the four dihedral angles. Write this data in a table.

Q8:   Plot the Time Series of the four dihedral angles (angle as a function of time).

Q9:   Based on your finding in **Q7** and **Q8**, do you think the job you just ran did a good job at sampling the molecule's conformation space?

B. SOLUTION SIMULATION

In this section you will simulate the dynamics of a solvated molecule. For simplicity and computational speed we will study a single methanol molecule (the solute) in a sphere of explicit water molecules (the solvent). Simulations of large solvated molecules are done in a similar way. To expedite the lab you are provided with the necessary CHARMM command files (in the *lab5* directory). These files are also printed out at the end of this chapter.

The simulation is divided into four steps. In each part you are expected to read and understand the (given) CHARMM command files, execute the simulation and answer the questions.

1) Model Generation - *MEOH_prep.inp*

Our model system includes a single methanol (CH3OH) molecule surrounded by a 7Å sphere of TIP3P water molecules (see details in the 'background' section). No boundary condition is imposed on this water 'droplet'. As you will see, the water droplet retains its integrity during the short done in this lab. For longer time simulation a boundary condition must be imposed on the water 'droplet'.

WATER
SPHERE

The system is constructed in the following way. First the methanol molecule is generated and its carbon atom is placed at the origin of the axes system. Then a 7Å sphere of water is overlaid on it (the center of the sphere also at the origin of the axes system). This water sphere has the density of bulk water (in fact it was cut out of a larger water sphere which represents bulk water). Finally the water molecules that *overlap* the MeOH are removed from the system (for this purpose water molecules are represented by 2.8Å spheres and overlap is checked with respect to MeOH's heavy atoms).

- Read the *MEOH_prep.inp* file. Be sure that you recognize the different steps in the model generation process.
- Run the model preparation simulation:

    charmm < *MEOH_prep.inp* > *MEOH_prep.out*

Q10:   How many water molecules are in the 7Å 'droplet'? How many water molecules were removed due to overlap with the solute? (look for the answers in the output file)

- Start QUANTA (if you exited it earlier).
- **Import** the coordinates of the solvated methanol *meoh_sol.crd*.
- Use color and display options to see the solute in the solvent. Use Solid Models to see a space-filling picture.

2) Initialization - *MEOH_init.inp*

The *MEOH_init.inp* command file performs the initialization phase.

Q11:   Read through the command file and specify what steps are taken in the initialization phase (minimization, dynamics etc.). Specify what algorithms are used, how many steps, size of time step, length of dynamics, heat up rate, simulation temperature.

- Run the initialization simulation:

   charmm < *MEOH_init.inp* > *MEOH_init.out*

Q12:   Write down the total energy before and after the minimization.

Q13:   Compare the radius of gyration of the water droplet (before & after minimization and after dynamics) and the heavy atom RMS distance from the initial structure (after minimization and after dynamics). Based on this data, what can you deduce about structural changes in the system?

- Extract the average energies that were calculated every 200 steps (0.2 ps), use the **grep** command to do so. Direct the results to a file for future use.

   grep 'AVER>' *MEOH_init.out* > *<grep_output>*

Q14:   Plot the graphs of the Total Energy, Potential energy and Temperature as a function of time.

- Check the stability of the last simulation by calculating the $\log_{10}(<\Delta E>/<E>)$ value over the last 4 ps of simulation.

Q15:   Based of the fluctuations of the total energy, what is the quality of the equilibration part of the simulation?

3) Production Run - *MEOH_run.inp*

In this part the dynamics of the 'equilibrated' system (in real studies, the equilibration part should take much longer) is simulated for an additional 5 ps.

- Run the production simulation:

  charmm < *MEOH_run.inp* > *MEOH_run.out*

Q16:   Check the radius of gyration and RMS distance at the end of
                the production run. Compare them with those obtained
                during the initialization (**Q13**). Any changes?

- Extract the average energies that were calculated every 200 steps:

  grep 'AVER>' *RGDS_run.out* > *<grep_output>*

- Check the stability of the simulation by calculating the
  $\log_{10}(<\Delta E>/<E>)$ value over the whole 5 ps of simulation (starting at
  time t=6 ps).

Q17:   Record the result and compare it to that of **Q15**.

- Use QUANTA to view the trajectory. Goto **Application/Dynamics
  Animation**, then **Select Trajectory** and **Open** the *meoh_sol.trj*
  trajectory file (generated during the production run). Then **Set Up
  Animation**, choose clock speed 50 and **CREATE**. Use the
  Clock/Cycle options to view the motion. When done **Exit Dynamics
  Animation**.

- **Exit** QUANTA.

4) Analysis - *MEOH_anal.inp*

Finally the production trajectory is to be analyzed. As usual, there are
many possible ways to analyze a trajectory and many different observable
that can be calculated. In this exercise we will focus on the following
question: "Does the water environment around the solute change during
the dynamics?" We would also like to know whether similar changes (if
any) occur at the hydroxyl end of the molecule and at the methyl end.

To answer this question (in a rather simplistic way) we will check, for
each frame in the trajectory, which water molecules are within a given

distance from the methanol's O atom and which water molecules are within a given distance from the methanol's C atom.

Unlike previous CHARMM command files that you have seen, this analysis procedure is fine-tuned through input from the user. There are two types of inputs that the program expects:

type O or C   (the MeOH atom which is the focus of the analysis)
dist <a number>   (the distance from the MeOH that defines the
                              neighborhood)

Upon execution you will use **grep** to filter the output and printout only the relevant lines. To execute type:

charmm type=<type> **dist**=<num> < *MEOH_anal.inp* | **grep** OH2 > *output*

The following example will print frame by frame the identity of the water molecules that are within 3.3Å from the solute's O atom. The output is directed into the *output* file.

charmm type=O **dist**=3.3 < *MEOH_anal.inp* | **grep** OH2 > *output*

The ID numbers of the water molecules are in the column before last.

Q18:   Analyze the solvated methanol trajectory.
   i)   Choose two distances around the O atom (e.g., 2.8Å and 3.3Å) and two distances around the C atom (they have to be slightly larger). In each of the four cases list the ID numbers of the water molecules frame by frame. (HINT: You can play with the distance parameter for best results).
   ii)   How many neighboring water molecules do the MeOH O and C atoms have at each distance?
   iii) Based on these results, discuss our original question: "Does the water environment around the solute change during the dynamics?"

IV. LAB WRITE-UP

Answer all the questions that were posed during the lab.

MEOH_prep.inp

```
* Lab 5: Preparation of solvated MeOH
*

OPEN READ CARD UNIT 1 NAME MODEL.TOP
READ RTF   CARD UNIT 1
OPEN READ CARD UNIT 2 NAME MODEL.PAR
READ PARAMeter CARD UNIT 2

!=============================
! Setup the Methanol molecule
!=============================

! Generate methanol
READ SEQUance CARD
* MEOH
*
1
MEOH
GENErate MEOH SETUp

! build methanol coordinates based on the parameters
IC PARAmeter
IC SEED 1 CB      1 OG      1 HG1
IC BUILd

!======================================
! Setup the surrounding water molecules
!======================================

! get sequence of water sphere from a file
OPEN UNIT 1 READ FORMatted NAME watsph7.crd
READ SEQUence UNIT 1 COOR
REWInd UNIT 1

! generate a new segment of water molecules
GENErate WAT NOANgle NODIhedral

! get the coordinates of the water molecules
READ COORdinate CARD UNIT 1 APPEnd
CLOSe UNIT 1

!=================================================
! remove any water molecule that overlaps the MeOH
! (mimic water by a 2.8 A spheres)
!=================================================

! { delete waters which overlap with protein }
DELEte ATOM SORT -
          SELEect .BYREs. (TYPE OH2 .AND. -
          ((TYPE CG .OR. TYPE OG) .AROUnd. 2.8)) END
```

```
!================================
! write out psf and coordinates
!================================

! write PSF
OPEN WRITe CARD UNIT 3 NAME meoh_sol.psf
WRITe PSF  CARD UNIT 3

! write crd
OPEN WRITe CARD UNIT 4 NAME meoh_sol.crd
WRITe COORdinate CARD UNIT 4

STOP
```

MEOH init.inp

```
* Lab 5: Initialization of solvated MeOH
*

! open topology and parameter
OPEN READ CARD UNIT 1 NAME MODEL.TOP
READ RTF   CARD UNIT 1
OPEN READ CARD UNIT 2 NAME MODEL.PAR
READ PARAMeter CARD UNIT 2

!====================================
! Setup the Methanol-Solvent system
!====================================

! read PSF
OPEN READ CARD UNIT 3 NAME meoh_sol.psf
READ PSF   CARD UNIT 3

! read initial crd
OPEN READ CARD UNIT 4 NAME meoh_sol.crd
READ COORdinate CARD UNIT 4

! non bonded interactions
NBONd CUTNb 150.0 CTONnb 145.0 CTOFnb 149.0 SHIFt VSWItch -
      CDIElectric EPSIlon 1.0

! save a copy of the initial coordinates
COORdinate COPY COMParison

!=========================================================
! Check radius of gyration of original water droplet
!=========================================================

COOR RGYR SELEct (RESName TIP3 ) END

!===========================
! Mimimize initial structure
!===========================

ENERgy
MINImize SD   NSTEp  500 NPRInt 50
ENERgy

!==================================
! check changes after minimization
!==================================

COOR ORIENT RMS SELECT (.NOT. HYDROGEN) END
COOR RGYR SELEct (RESName TIP3 ) END

!=================
! initial heat up
!=================

! open the restart file
OPEN WRITe FORMatted UNIT 10 NAME meoh_sol.res

! heat the system from 0 K to 300 K
DYNAmics VERLet STRT -
        NSTEp  3000  TIMEstep 0.001 -
```

```
        FIRST    0.0  FINAlT    300.0  TEMInc   75.0 -
        IHTFrq   500  NTRFrq    1000 -
        INBFrq     0  IHBFrq      -1 -
        NPRInt    50  IPRFrq     200 -
        IASORS     1  IASVEL       1 -
        IUNWrite 10

!=======================
! initial equilibration
!=======================

! Close restart file and reopen it
CLOSe UNIT 10
OPEN   READ  FORMatted UNIT 10 NAME meoh_sol.res
OPEN   WRITe FORMatted UNIT 11 NAME meoh_sol.res1

! equilibrate the system for 3 ps
DYNAmics VERLet RESTart -
        NSTEp 3000  TIMEstep 0.001 -
        NPRInt  50  IPRFrq      200 -
        IHTFrq   0  NTRFrq     1000 -
        INBFrq   0  IHBFrq       -1 -
        IUNRead 10  IUNWrite 11

! write coordinates
OPEN WRITe CARD UNIT 15 NAME meoh_sol_1.crd
WRITe COORdinate CARD UNIT 15

!======================================
! check changes after initial dynamics
!======================================

COOR ORIENT RMS SELECT (.NOT. HYDROGEN) END
COOR RGYR SELEct (RESName TIP3 ) END

STOP
```

MEOH_run.inp

```
* Lab 5: Production run of solvated MeOH
*

OPEN READ CARD UNIT 1 NAME MODEL.TOP
READ RTF   CARD UNIT 1
OPEN READ CARD UNIT 2 NAME MODEL.PAR
READ PARAMeter CARD UNIT 2

!===================================
! Setup the Methanol-Solvent system
!===================================

! read PSF
OPEN READ CARD UNIT 3 NAME meoh_sol.psf
READ PSF   CARD UNIT 3

! read initial crd
OPEN READ CARD UNIT 4 NAME meoh_sol.crd
READ COORdinate CARD UNIT 4

! non bonded interactions
NBONd CUTNb 150.0 CTONnb 145.0 CTOFnb 149.0 SHIFt VSWItch -
      CDIElectric EPSIlon 1.0

! save a copy of the initial coordinates
COORdinate COPY COMParison

!=================================
! run production dynamics for 5 ps
!=================================

! open the restart file
OPEN READ  FORMatted UNIT 10 NAME meoh_sol.res1
OPEN WRITe FORMatted UNIT 11 NAME meoh_sol.res2

! open trajectory file
OPEN WRITe UNFOrmatted UNIT 12 NAME meoh_sol.trj

! run production simulation for 5 ps
DYNAmics VERLet RESTart -
      NSTEp  5000  TIMEstep 0.001 -
      NPRInt   50  IPRFrq    200 -
      IHTFrq    0  NTRFrq      0 -
      INBFrq    0  IHBFrq     -1 -
      NSAVC   200  IUNCRD     12 -
      IUNRead  10  IUNWrite   11

!=================================
! check changes after minimization
!=================================

COOR ORIENT RMS SELECT (.NOT. HYDROGEN) END
COOR RGYR SELEct (RESName TIP3 ) END


END
```

MEOH_anal.inp

```
* Lab 5: Analysis of solvated MeOH production run
* a passed variable (type) for CG or OG
* a passed variable (dist) for distance
*

OPEN READ CARD UNIT 1 NAME MODEL.TOP
READ RTF   CARD UNIT 1
OPEN READ CARD UNIT 2 NAME MODEL.PAR
READ PARAMeter CARD UNIT 2

! arrange type variable
IF @type .EQ. O  SET typ1 OG
IF @type .EQ. C  SET typ1 CB

!==================================
! Setup the Methanol-Solvent system
!==================================

! read PSF
OPEN READ CARD UNIT 3 NAME meoh_sol.psf
READ PSF   CARD UNIT 3

! read initial crd
OPEN READ CARD UNIT 4 NAME meoh_sol.crd
READ COORdinate CARD UNIT 4

! non bonded interactions
NBONd CUTNb 150.0 CTONnb 145.0 CTOFnb 149.0 SHIFt VSWItch -
      CDIElectric EPSIlon 1.0

!=============================
! analyze production dynamics
!=============================

OPEN UNIT 10 READ UNFORmatted NAME meoh_sol.trj

TRAJectory IREAd 10 NREAd 1 SKIP 200

SET frame 1
LABEl get_frame

TITLe
* <OH2> frame number: @frame
*

    TRAJectory READ

    PRINT COORdinates SELECT (TYPE OH2 .AND. -
            (TYPE @typ1 .AROUnd. @dist )) END

! close the loop
INCRement frame
IF frame LE 20 GOTO get_frame

END
```

Lab 6:   Molecular Dynamics and Analysis

I. OBJECTIVE

This lab addresses Molecular Dynamics (MD) simulation on a more advanced level than the previous lab. In this lab you will construct a molecular model of a complex biomolecule – the protein myoglobin with a CO ligand bound to it (MbCO). You will perform a MD simulation of this molecule and apply several analysis tools to extract information from the resulting dynamics trajectory.

Because of the size of the system, several simplifications will be introduced to make the execution of this lab feasible. Specifically, the 'polar-hydrogen representation' will be used and the simulation will be performed in vacuum (with distance dependent dielectrics). In the 'polar-hydrogen representation' only polar hydrogens are explicitly included in the model; all other non-polar hydrogens are schematically included in an 'extended' version of their neighboring heavy atom (e.g., CH_3 is represented by a single 'united' carbon atom of mass 15). In addition, you will have to determine by yourself which of the two possible histidine configurations to choose for the different histidine residues in the molecule.

This is also the first lab in which you are supplied with *CHARMM Template Files* for your convenience (see instructions in the introduction to this *Guide*).

II. BACKGROUND

The basic concepts of molecular dynamics simulations were discussed in Lab 5 along with the practical stages involved in executing such a simulation. The present lab builds on that and introduces a few additional tools, among them using the SHAKE algorithm to constrain bonds and a few more analysis tools.

SHAKE

One well-known limitation of molecular dynamics simulations is the need to use very short time steps during the integration of the equation of motion. For proteins and other organic molecules this time step is on the order of 0.5 fs (0.5×10^{-15} s). A consequence of the small time steps is that it is hard to propagate the system to long times. Indeed, typical protein simulations are limited today to the nanosecond time range.

The origin for this limitation is that the time step must be small enough so that even the fastest motions in the protein will be sampled ten to twenty times during their vibration periods. Without fulfilling this requirement the numerical integration will not be stable. For organic molecules, the fastest motions involve stretching the bonds to hydrogen atoms (e.g., C-H at ~3000 cm^{-1}) and in second place all other bond-stretching motions (e.g., C=O at ~1800-1500 cm^{-1}). If these stretching motions could be ignored or somehow represented in a different way, than a longer time step could be used (leading to longer simulation times for the same computational effort).

SHAKE   SHAKE is just this kind of an algorithm [W. F. van Gunsteren and H. J. C. Berendsen, *Mol. Phys.* **34:** 1311 (1977); see also: W. F. van Gunsteren and M. Karplus, *Macromolecules* **15:** 1582 (1982)]. What SHAKE does is to constrain bond-lengths at their ideal values so that their stretching motions do not have to be numerically integrated.

Generally, the constraint on the bond between atoms i and j is expressed as

$$r_{ij}^2 - d_{ij}^2 = 0$$

where r_{ij} is the actual distance between the atoms and d_{ij} is the ideal value. In a numerical simulation this relation is not generally true, so it is said that the constraint is satisfied whenever the relative deviation is sufficiently small, i.e.,

$$s_{ij} = (r_{ij}^2 - d_{ij}^2)/d_{ij}^2 < \varepsilon$$

where ε is a specified tolerance. SHAKE iteratively adjusts the atomic positions after each time step in order to simultaneously satisfy all the constraints; i.e., until s_{ij} is smaller than ε for all the constrained bonds. As this procedure is much faster than actual integration, a significant saving in computer time is achieved.

In this lab you will apply SHAKE to the bonds involving hydrogen atoms, thus allowing a factor 2 increase in the time step, from 0.5 fs to 1 fs. This is the most common application of SHAKE.

Analysis Tools

In the previous labs you used one analysis tool - the time series - to follow the motion of specific internal coordinates. In this lab two more analysis tools will be used: the Internal Coordinates (IC) table and the time autocorrelation function.

Internal Coordinates table (IC table)

When analyzing the results of a dynamics simulation one is often interested in characterizing the motions of specific internal coordinates (e.g., dihedral angles). The IC table is used for extracting average values of internal coordinates as well as their fluctuations.

IC TABLE

The IC table has five columns for each set of four atoms I-J-K-L that define a proper or improper dihedral angle. In case of a proper dihedral the five columns of the table hold the following values: R (I-J), θ (I-J-K), Φ (I-J-K-L), θ (J-K-L), R (K-L), where R are bond distances, θ are bond-angles and Φ is the dihedral. In case of an improper dihedral (indicated by an asterisk) these columns are: R (I-K), θ (I-K-J), Φ (I-J-K-L), θ (J-K-L), R (K-L).

Following is a sample IC table:

N	I	J	K	L	R(I(J/K))	θ(I(JK/KJ)	Φ(IJKL)	θ(JKL)	R(KL)
1	2 N	1 CL	1 *C	1 O	1.33	117.50	180.00	121.50	1.23
2	1 CL	1 C	2 N	2 CA	1.52	117.50	180.00	120.00	1.49

Time autocorrelation function

A powerful group of analysis tools are the time autocorrelation functions. For any time dependent property $A(t)$, such as bond length or dihedral angle, the time autocorrelation function $C_A(t)$ is defined as

**AUTO-
CORRELATION
FUNCTION**

$$C_A(t) = <A(t) \cdot A(0)>$$

This function measures the correlation of the property $A(t)$ to itself at two different times, separated by the time interval t, averaged over the whole trajectory. The autocorrelation function is *even* in time (i.e., $C_A(t) = C_A(-t)$) and it is *stationary* (i.e., $<A(t + \tau)A(\tau)> = <A(t) A(0) >$).

An important property of the time autocorrelation function $C_A(t)$ is that by taking its Fourier Transform, $F\{C_A(t)\}_\omega$, one gets a spectral decomposition of all the frequencies that contribute to the motion. For example consider the motion of a single particle in a harmonic potential (harmonic oscillator). The 'time series' describing the position of the particle as a function of time is given by the $cos(\omega_0 t)$ function (Figure I). The autocorrelation function is given by a *cos* function with a period $2\pi/\omega_0$ (Figure II), and its Fourier Transform gives a spectrum with a single sharp peak at ω_0 (Figure III). The resulting frequency can be used to extract the (real or effective) local force constant, since $K_0 = m\omega_0^2$.

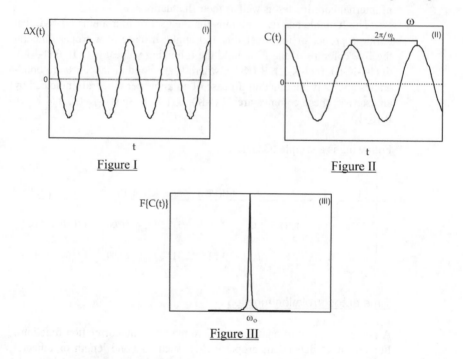

Figure I   Figure II

Figure III

III. PROCEDURE

Create a new sub-directory *lab6* and change to that directory. Copy the content of the *$Lab/lab6* directory to your own directory (**cp -r** *$Lab/lab6/** **.**) [the **-r** flag indicates that sub-directories under *lab6* will also be copied]. You now have several CHARMM template (*.tmpl*) and stream (*.str*) files and a sub-directory named *lib*. View the content of this sub-directory (**ls** *lib*) and identify the topology, parameter and coordinate files you will be using. Now create a new sub-directory *out* within your *lab6* directory (**mkdir** *out*). You will direct the results of your calculations into this *out* directory.

Throughout this lab you will be given CHARMM *template* files to modify. Refer to the **CHARMM Template Files** instructions in the introduction to this *Guide*. These template files are also printed out at the end of the Lab 6 chapter.

You will also be using CHARMM stream files (*.str*) in this lab. Recall that these files include groups of often-repeated CHARMM commands, which at runtime are included in the main command sequence using the STREam keyword. See the **CHARMM Primer** for details.

A. SYSTEM INITIALIZATION

1)   Generate PSF - *gen_psf.tmpl*

The first stage is to generate a PSF for the molecular complex. Recall that you are using the 'polar-hydrogen representation', which is defined by the CHARMM-19 topology and parameter set. In this case the MbCO complex is constructed from three molecular entities (myoglobin, the heme group and the CO ligand). These molecules are PATChed together (the PATChes are defined in the topology file). After generating the PSF and reading the PDB X-ray crystal coordinates (*lib/mbco_pdb4.crd*) the polar-hydrogens are built into the model. These hydrogens are then minimized and a new PSF and coordinate files are written (to what directory?).

MODIFY: Practice your SELEct skills - You are asked to modify the command line that counts the number of heavy protein atoms to give the number of backbone atoms and number of heavy Heme atoms.

After you have modified the template file and created the appropriate input file, execute the job

      charmm < *gen_psf.inp* > *out/gen_psf.out* &

you can see the number of atoms you calculated using the UNIX **grep** command,

grep 'MY RESULTS:' *out/gen_psf.out*

NEW CHARMM COMMANDS:
 PATCh DEFIne SELEct HBUIld   CONStraint

Q1:   Record your results. How many atoms of the different types
                  are there in the system?

2) <u>Check Histidine Conformation</u> - *check_his.tmpl*

One of the many problems associated with structure determination is the
conformation (and protonation state) of the histidine residues in the
molecule. We have 12 histidines at residue numbers 12, 24, 36, 48, 64,
81, 82, 97, 113, 116, and 119 (the proximal His 93 was intentionally left
out since it is directly bound to the heme). To probe which of the two
conformations (H-Nδ1 or H-Nϵ2; see schematic drawings associated with
Q3 below) is more stable for each histidine, we initiate the two
conformations, minimize the amino acids which are within 6 Å from the
histidine, and then determine the interaction energy of that histidine with
the rest of the molecule. The *check_his.tmpl* script has to be modified
slightly and run separately for each histidine residue.

TIP: For clarity and convenience, from now on the topology,
                  parameter, PSF and coordinate files will be read using the
                  two STREam files, *load_psf.str* and *load_crd.str*. Make
                  sure you understand the content of these two files. Refer to
                  the ***CHARMM Primer*** if you don't remember what
                  STREam files are.

MODIFY: First define the names of the setup files and then set the
                  value of *h* to the residue you are studying. This *h* variable
                  has to be changed for each residue number.

First set *h* to 12, and run

          charmm < *check_his.inp* > *out/H12.out* &

then change *h* to 24, and run

          charmm < *check_his.inp* > *out/H24.out* &

... up to setting *h* to 119 and running

To conveniently tabulate the results, after all 11 residues have been examined by CHARMM, type

   ./getHisInte

```
NEW CHARMM COMMANDS:
  STREam  INTEraction  COOR  COPY
```

Q2:   Write a table with the interaction energies of the two histidine-conformations (H-Nδ1 and H-Nε2) for the 11 histidine residues you tested. Which are the <u>four</u> residues that are more stable in the H-Nε2 conformation?

"Hε2 not seen by neutron diffraction"   "Hδ1 less stable by calculations"

Q3:   Briefly describe why the computed H64 preference is H-Nε2 (given your results), yet neutron scattering results show the absence of a hydrogen bond with the carbon monoxide ligand, indicating that H-Nδ1 is the proper conformation of H64. Remember that the computations are in vacuum; what about the dielectric constant?

3)   <u>Update the PSF</u> - *update_psf.tmpl*

After changing the conformation from H-Nδ1 to H-Nε2 for several (four!) histidine residues (according to the results of the previous section) regenerate the PSF and reassign the hydrogen coordinates.

MODIFY: The names of the setup files and the PATChes for changing the conformations of the histidines. Make sure you know the names of the new PSF and CRD files.

To execute the update job type:

   charmm < *update_psf.inp* > *out/update_psf.out* &

B. MOLECULAR DYNAMICS

When performing Molecular Dynamics simulations of proteins one must first prepare the system (assign velocities, heat up to the desired temperature and equilibrate). Only then a 'production' simulation can be run. In this section you will perform these two stages in a way similar to real research, with the sole difference that the simulation times will be much shorter than those used for real research.

   In these simulations you will restrain the motion of the hydrogen atoms using the SHAKE algorithm and set the integration time step to 1 fs (0.001 ps).

1)   Preparatory Stage - *mbco_prep.tmpl*

The preparatory stage of the dynamics is divided into three steps:
(i)   Assign velocities from a Maxwellian distribution and run for 0.2 ps.
(ii) Run 0.8 ps of dynamics, checking and rescaling the velocities every 100 steps.
(iii) Run 1.0 ps of equilibration (without velocity rescaling)

MODIFY: The names of the setup files (use the updated PSF and CRD files). Make sure you are familiar with the various output files.

Running this job may take some time. To use your time more efficiently continue to the next stage and then execute both stages together.

2)   Production Stage - *mbco_run.tmpl*

After equilibrating the system you can carry out the production dynamics. Typically this will last for at least 50 ps, but to save time we will limit the exercise only to 2 ps of simulation time.

MODIFY: The names of the setup files and set the appropriate keywords in the DYNAmics command to perform 2.0 ps of simulation using 1 fs time steps, with no velocity rescaling (NSTEp, TIMEstep, IEQFrq).

This stage of the exercise requires a few minutes of computer time. To execute both stages, one after the other, you are supplied with a small UNIX script file that will submit the jobs for you (take a look at this script file). To run it simply type,

    runjobs &

You can now leave the computer and return when the calculation is done.

NOTE: The *runjobs* script-file simply submits the second job (*mbco_run.inp*) immediately after the first one (*mbco_prep.inp*) is finished; i.e., it is equivalent to the following two lines,

    charmm < *mbco_prep.inp* > *out/mbco_prep.out* &
    charmm < *mbco_run.inp* > *out/mbco_run.out* &

```
NEW CHARMM COMMANDS:
    SHAKe
```

C. ANALYSIS OF A DYNAMICS TRAJECTORY

When analyzing the results of a dynamics simulation one is attempting to characterize the motions that occurred. Since there are many ways to characterize such motions we shall introduce three techniques in this lab: (1) Average structure and atomic fluctuations, (2) Average values of internal coordinates and their fluctuations, and (3) Time correlation functions and their corresponding spectrum.

1) Average structure and atomic fluctuations - *anal1.tmpl*

In this type of analysis coordinate-sets from the dynamics trajectory are averaged together. The average value and the fluctuations around the average are analyzed. The commands in the *anal1.tmpl* file read the trajectory generated by the 2 ps production run and averages it (using the command COORdinate DYNAmics). This averaging procedure puts the values of the atomic fluctuations in the fourth column of the main coordinate set, called WMAIN. These values are then manipulated to give fluctuations for specific types of atoms and specific segments: (a) all non-H atoms, (b) all carbon atoms, (c) only C_α atoms, (d) only C_β atoms, (e) all N atoms, and (f) all O atoms, for the whole molecule and for the Heme group.

TIP: Each coordinate set in CHARMM has <u>four</u> components (columns). The first three are the Cartesian X, Y and Z coordinates while the fourth is a "weight" variable. This "weight" column is used in averaging and in analysis and can reflect different properties at different times. The fourth column of the main coordinate set is called WMAIN ("**W**eight for **MAIN** coordinates).

MODIFY: The names of the setup files.

To execute the simulation type:

     charmm < *anal1.inp* > *out/anal1.out* &

NEW CHARMM COMMANDS:
        COOR DYNAmics   SCALar

Q4:   Extract the data on RMS fluctuations of the various atom types in the different segments from the *anal1* output file. Present the results in a table and explain the trends you see (e.g., all C vs. C_α vs. C_β).

2) <u>Averages and fluctuations of IC</u> - *anal2.tmpl*

Now you will focus on several internal coordinates of special interest and study their dynamics. To focus on key internal coordinates (such as dihedral angles that can indicate ring flips) you first have to generate an IC Table. Then the trajectory can be read and the average values of the specified internal coordinates along with their fluctuations can be calculated. The structure of the IC table is explained in the **CHARMM Primer** and in the 'Background' notes to this lab. The template file you'll be using to perform these tasks is *anal2.tmpl*.

The eight internal coordinates defined in this template file characterize: (a) the back-bone near His 64 (three dihedral angles), (b) His 64 ring motion (one dihedral angle), (c) the back-bone near His 93 (three dihedral angles), and (d) Tyr 103 ring motion (one dihedral angle). You are required to <u>add</u> another (ninth!) internal coordinate, one that will characterize the ring motion of His 93.

MODIFY: The names of the setup files and add the definition of an internal coordinate that will characterize the ring motion of His 93 (use the definition for His 64 as a working sample).

To perform the simulation type:

charmm < *anal2.inp* > *out/anal2.out* &

NEW CHARMM COMMANDS:
    IC EDIT   IC FILL   IC DYNAmics   IC PRINt

Q5: Extract the two IC Tables (with averages and with fluctuations) from the *anal2* output file. Which of these two Tables is more informative in terms of dynamics? Describe the general differences between the dynamics of the backbone regions and that of the rings?

3) <u>Time autocorrelation functions and spectra</u> - *anal3.tmpl*

In this exercise you will analyze the CO bond stretching motion using its time autocorrelation function. The job that you will run first extracts the 'time series' of the CO bond stretching, then an autocorrelation function is calculated from this time-series and finally the Fourier Transformed spectrum of this function will be calculated.

NOTE: The data set that you are analyzing does not include enough points to generate a spectrum in the relevant frequency range. Thus, the "spectrum" you will get is essentially meaningless. A much longer trajectory is needed for a real analysis of this motion. You are, however, calculating the 'spectrum' for the sake of completeness and acquaintance with the procedure.

The template file you'll be using to perform this task is *anal3.tmpl*. This template file includes several CHARMM commands from its

CORRelations module: MANTime (gives the simple time series), CORFUNction (gives the time autocorrelation function) and SPECTrum (gives the spectrum).

✳   **MODIFY:**   The names of the setup files. Make sure that you recognize the different output files generated by the program.

To perform the simulation type:

charmm < *anal3.inp* > *out/anal3.out* &

NEW CHARMM COMMANDS:
   CORRELation

Q6:   Use the files written by *anal3.inp* to generate two plots: (a) the time series of the CO bond (bond length vs. time), (b) its correlation function (what are the bounds of this function?). What can you learn from these plots regarding the CO stretching motion?

IV. LAB WRITE-UP

Answer all the questions that were posed during the lab.

gen_psf.tmpl

```
* Lab 6: MbCO Dynamics and Analysis
* Generate PSF of Sperm Whale MbCO from X-ray (PDB) at 260 K.
*

BOMLev -1

! Read Topology and Parameter files
! -------------------------------
SET top lib/toph19.inp
OPEN UNIT 1 CARD READ NAME @top
READ RTF CARD UNIT 1
CLOSe UNIT 1

SET par lib/param19.inp
OPEN UNIT 1 FORMatted READ NAME @par
READ PARAmeter CARD UNIT 1
CLOSe UNIT 1

! Generate PSF.
! =======================================
! MbCO: a. 153 residue polypeptide Mb
!       b. Heme group bound to His 93 NE2
!       c. CO liganding heme iron atom
! =======================================


READ SEQUence CARDs            ! Read and generate the Myoglobin
* Sperm-whale MYOGLOBIN sequence
*
   153
VAL LEU SER GLU GLY GLU TRP GLN LEU VAL LEU HIS VAL
TRP ALA LYS VAL GLU ALA ASP VAL ALA GLY HIS GLY GLN
ASP ILE LEU ILE ARG LEU PHE LYS SER HIS PRO GLU THR
LEU GLU LYS PHE ASP ARG PHE LYS HIS LEU LYS THR GLU
ALA GLU MET LYS ALA SER GLU ASP LEU LYS LYS HIS GLY
VAL THR VAL LEU THR ALA LEU GLY ALA ILE LEU LYS LYS
LYS GLY HIS HIS GLU ALA GLU LEU LYS PRO LEU ALA GLN
SER HIS ALA THR LYS HIS LYS ILE PRO ILE LYS TYR LEU
GLU PHE ILE SER GLU ALA ILE ILE HIS VAL LEU HIS SER
ARG HIS PRO GLY asp PHE GLY ALA ASP ALA GLN GLY ALA
MET ASN LYS ALA LEU GLU LEU PHE ARG LYS ASP ILE ALA
ALA LYS TYR LYS GLU LEU GLY TYR GLN GLY

GENErate 1MBC SETUp NODIhedral

READ SEQUence CARDs            ! Read and generate the HEME group
```

```
* HEME GROUP
*
   1
HEME

GENErate HEME SETUp NODIhedral

PATCh PHEM 1MBC 93 HEME 1      ! Attach proximal H93 to the HEME

READ SEQUence CARDs            ! Read and generate the CO model
* CO MOLECULE
*
   1
CO2P

GENErate CO2P WARNing SETUp NODIhedral

PATCh PLIG CO2P 1 HEME 1 1MBC 93      ! Attach the ligand CO2P to the heme

! Coordinates.
! -----------
SET crd lib/mbco_pdb4.crd
OPEN UNIT 1 READ CARD NAME @crd
READ COORdinate PDB UNIT 1
CLOSe UNIT 1


! ********************************************************
! * MODIFY
! * Change the following comands to calculate:
! *     (1) number of heavy atoms
! *     (2) number of backbone atoms (type C, CA, N)
! *     (3) number of heavy HEME atoms.
! ********************************************************

! Find number of heavy protein atoms
! -------------------------------------
DEFIne count1 SELEct (SEGId 1MBC .AND. .NOT. HYDRogen) END
set 1 ?NSEL

! Find number backbone atoms
! --------------------------
DEFIne count2 ???
SET 2 ?NSEL

! Find number of heavy HEME atoms
! -------------------------------
DEFIne count3 ???
SET 3 ?NSEL
```

```
! Print the result
! ---------------
TITLe
* MY RESULTS: heavy atoms    = @1
* MY RESULTS: backbone atoms = @2
* MY RESULTS: heavy heme atoms = @3
*


!===================================================================
! Place Hydrogens (First guess, second in the field of other H)
!===================================================================
HBUIld SELEct (HYDRogen .AND. .NOT. INITial) END
HBUIld SELEct (HYDRogen) END

! Non-bonded specifications
! ------------------------
NBONd -
    INBFrq 25 CUTNb 10.0 CTONnb 6.5 CTOFnb 9.0 -
    E14Fac 1.0 RDIElectric VSWITch  SHIFt

! Constrain all non-hydrogen atoms
! -------------------------------
CONStraint FIX SELEct (.NOT. HYDRogen) END

! Few steps of minimization hydrogens ONLY.
! ----------------------------------------
MINImize SD  NSTEp 45 NPRInt 5
MINImize ABNR NSTEp 5 NPRInt 1

! Write New Coordinates.
! ---------------------
SET g lib/mbco.crd
OPEN UNIT 1 WRITe FORMatted NAME @g
WRITe COORdinate CARD UNIT 1
* COORDINATES: Sperm Whale MbCO from X-ray (PDB) at 260 K.
* Parameters and Topology files: param19.inp, toph19.inp
*

! Write the PSF for MbCO.
! ----------------------
SET s lib/mbco.psf
OPEN UNIT 1 CARD WRITe NAME @s
WRITe PSF CARD UNIT 1
* PSF: Sperm Whale MbCO from X-ray (PDB) at 260 K.
* Parameters and Topology files: param19.inp, toph19.inp
*

STOP
```

check_his.tmpl

```
* Lab 6: MbCO Dynamics and Analysis
* Check for stable conformations of Histidines
*

BOMLev -1
FASTer on

! ***********
! * MODIFY *
! ***********
set h 12   ! Change to 12,24,36,48,64,81,82,97,113,116, and 119.

! ************************
! * MODIFY the file names
! ************************
SET t ???                       ! topology file
SET p ???                       ! parameter file
SET s ???                       ! PSF
SET c ???                       ! coordinate file

! load the topology, parameter, psf and coordinate files
STREam load_psf.str
STREam load_crd.str

COOR COPY COMP        ! Retain coordinates for equilivant starting
                      ! points.

! =========================================================
! STEP A: Calculate interactions of HIS conformation Nd1
! =========================================================

! Constrain residues beyond 6 A of target histidine.
! -------------------------------------------------
CONStraint FIX SELEct (.NOT. (.BYRES. ((SEGId 1MBC .AND. RESId @h) -
        .AROUND. 6)) .OR. SEGId CO2P) END

! Non-bonded specifications
! -------------------------
NBONd -
        INBFrq 25 CUTNb 10.0 CTONnb 6.5 CTOFnb 9.0 -
        E14Fac 1.0 RDIElectric VSWITch  SHIFt

! Few steps of minimization of residues within 6 A target.
! --------------------------------------------------------
MINImize SD  NSTEP 45 NPRINT 5
MINImize ABNR NSTEP 5 NPRINT 1
```

```
! Interaction energy of the histidine.
! ---------------------------------
INTEraction SELEct (SEGId 1MBC .AND. RESId @h) END -
     SELEct .NOT. (SEGId 1MBC .AND. RESId @h) END


! =====================================================
! STEP B: Calculate interactions of HIS conformation Ne2
! =====================================================

COOR COPY             ! Retrieve original coordinates.

PATCH HS2 1MBC @h     ! Patch Nd1 to Ne2 (prepare second configuration)

! Build the hydrogen and print hydrogen coordinate added.
! ------------------------------------------------------
PRINt COOR SELEct .NOT. init END
HBUIld
PRINt COOR SELEct .NOT. init END

! Constrain residues beyond 6 A of target histidine.
! --------------------------------------------------
CONStraint FIX SELEct NONE END
CONStraint FIX SELEct (.NOT. (.BYRES. ((SEGId 1MBC .AND. RESId @h) -
     .AROUND. 6)) .OR. SEGId CO2P) END

! Few steps of minimization.
! ----------------------
MINImize SD  NSTEP 45 NPRINT 5
MINImize ABNR NSTEP 5 NPRINT 1

! Interaction energy of the histidine.
! ---------------------------------
INTEraction SELEct (SEGId 1MBC .AND. RESId @h) END -
     SELEct .NOT. (SEGId 1MBC .AND. RESId @h) END

STOP
```

update_psf.tmpl

```
* Lab 6: MbCO Dynamics and Analysis
* Generate new PSF based of stable conformations of the Histidins
*

BOMLev -1
FASTer on

! ***********************
! * MODIFY the file names
! ***********************
SET t ???                      ! topology file
SET p ???                      ! parameter file
SET s ???                      ! PSF
SET c ???                      ! coordinate file

! load the topology, parameter, psf and coordinate files
STREam load_psf.str
STREam load_crd.str

! ***********************************************************************
! * MODIFY:
! * Set the prefered HIS configurations based on your results.
! * Enter a line for each His that is to be in the H-Ne2 conformation
! * (an example given for His 64).
! * HINT: You should endup with 4 such lines.
! ***********************************************************************

PATCH HS2 1MBC 64              ! Patch Nd1 to Ne2 for H64.
PATCH HS2 1MBC ??              ! Patch Nd1 to Ne2 for ???
PATCH HS2 1MBC ??              ! Patch Nd1 to Ne2 for ???
PATCH HS2 1MBC ??              ! Patch Nd1 to Ne2 for ???

! Build new hydrogens and print hydrogen coordinate added.
! ---------------------------------------------------------
PRINt COOR SELEct .NOT. init END
HBUIld
PRINt COOR SELEct .NOT. init END

! Write New Coordinates.
! ---------------------
SET g lib/mbco_h.crd
OPEN UNIT 1 WRITe FORMatted NAME @g
WRITe COORdinate CARD UNIT 1
* MbCO COORdinate file - with stable HIS conformations
*
CLOSe UNIT 1
```

```
! Write new PSF
! -------------
SET s lib/mbco_h.psf
OPEN UNIT 1 CARD WRITe NAME @s
WRITe PSF CARD UNIT 1
* MbCO PSF file - with stable HIS conformations
*
CLOSe UNIT 1

STOP
```

mbco_prep.tmpl

```
* Lab 6: MbCO Dynamics and Analysis
* Preperatory Molecular Dynamics of MbCO:
* Stage 1: 0.2 ps of dynamics with initial velocity assignment
* Stage 2: 0.8 ps dynamics with scaling of velovities
* Stage 3: 1.0 ps equilibration.
*

BOMLev -1

! ***********************
! * MODIFY the file names
! ***********************
SET t ???                        ! topology file
SET p ???                        ! parameter file
SET s ???                        ! PSF
SET c ???                        ! coordinate file

! load the topology, parameter, psf and coordinate files
STREam load_psf.str
STREam load_crd.str

! Non-bonded specifications
! -------------------------
NBONd -
   INBFrq 25 CUTNb 10.0 CTONnb 6.5 CTOFnb 9.0 -
   E14Fac 1.0 RDIElectric VSWITch  SHIFt

! Set constraints
! ----------------
CONStraint FIX SELE NONE END          ! No fixed atoms
SHAKe BONH TOLErance 1.0e-06 PARAmeter       ! SHAKE  bonds  to  Hydrogen
atoms


! =======================================
! Start dynamics: Preperatory MD stage 1
! =======================================
SET i out/mbco_prep1                              ! Current stage

OPEN UNIT 31 WRITe FORMatted  NAME @i.res    ! The RESTart file
OPEN UNIT 32 WRITe UNFOrmatted NAME @i.trj   ! The trajectory file
OPEN UNIT 33 WRITe FORMatted  NAME @i.ene    ! The energy output file

! Set Maxwellian distribution of velocities and run for 0.2 ps.
! -------------------------------------------------------------
DYNAmics LEAPfrog STRT -
   NSTEp 200  TIMEstep 0.001  -
   IPRFrq 100  NPRInt   50 FIRStt 300. FINAlt  300. -
```

```
     IASOrs  1   ISCVel   0 IASVel  1 ISEEd  456244 -
     IEQFrq  0   ICHEcw   0 -
     INBFrq  -1  IHBFrq   0 -
     IUNCrd  32  KUNIt    33 NSAVc   10 NSAVv   -1 -
     IUNWri  31  ISVFrq   200


! ========================================
! Restart dynamics: Preperatory MD stage 2
! ========================================
SET j @i                                    ! Previous stage
SET i out/mbco_prep2                         ! Current stage

OPEN UNIT 30 READ FORMatted  NAME @j.res     ! Previous RESTart file
OPEN UNIT 31 WRITe FORMatted  NAME @i.res    ! The RESTart file
OPEN UNIT 32 WRITe UNFOrmatted NAME @i.trj   ! The trajectory file
OPEN UNIT 33 WRITe FORMatted  NAME @i.ene    ! The energy output file


! Scale velocities for 0.8 ps
! --------------------------
DYNAmics LEAPfrog RESTart -
     NSTEp  800  TIMEstep 0.001  -
     IPRFrq 100  NPRInt   50 FINAlt 300.  -
     IASOrs 0    ISCVel   0 IASVel 1 -
     IEQFrq 100  ICHEcw   1 TWINdh 5. TWINdl -5. -
     INBFrq  -1  IHBFrq   0 -
     IUNCrd 32   KUNIt    33 NSAVc 10 NSAVv -1 -
     IUNRea 30   IUNWri   31 ISVFrq 200


! ==========================================
! Restart dynamics: Preperatory MD stage 3
! ==========================================
SET j @i                                    ! Previous stage
SET i out/mbco_prep3                         ! Current stage

OPEN UNIT 30 READ FORMatted  NAME @j.res     ! Previous RESTart file
OPEN UNIT 31 WRITe FORMatted  NAME @i.res    ! The RESTart file
OPEN UNIT 32 WRITe UNFOrmatted NAME @i.trj   ! The trajectory file
OPEN UNIT 33 WRITe FORMatted  NAME @i.ene    ! The energy output file


! Equilibration for 1 ps
! ----------------------
DYNAmics LEAPfrog RESTart -
     NSTEp 1000  TIMEstep 0.001  -
     IPRFrq 100  NPRInt   50 FINAlt 300.  -
     IASOrs 0    ISCVel   0 IASVel 1 -
     IEQFrq 0    ICHEcw   0 -
```

```
      INBFrq  -1  IHBFrq     0  -
      IUNCrd  32  KUNIt     33  NSAVc  10 NSAVv  -1 -
      IUNRea  30  IUNWri    31  ISVFrq 200


! Write the coordinates after Stage 3 of preparation
! ------------------------------------------------
OPEN UNIT 21 WRITe FORM NAME @i.crd
WRITe COOR CARDs UNIT 21
* MbCO: Coordinates after prperatory stage 3.
*
CLOSe UNIT 21


STOP
```

mbco_run.tmpl

```
* Lab 6: MbCO Dynamics and Analysis
* 2.0 ps Production Molecular Dynamics of MbCO.
*


BOMLev -1


! **********************
! * MODIFY the file names
! **********************
SET t ???                          ! topology file
SET p ???                          ! parameter file
SET s ???                          ! PSF
SET c ???                          ! coordinate file


! load the topology, parameter, psf and coordinate files
STREam load_psf.str
STREam load_crd.str


! Non-bonded specifications
! ------------------------
NBONd -
    INBFrq 25 CUTNb 10.0 CTONnb 6.5 CTOFnb 9.0 -
    E14Fac 1.0 RDIElectric VSWITch  SHIFt


! Set constraints
! ---------------
CONS FIX SELE NONE END                     ! No fixed atoms
SHAKe BONH TOLErance 1.0e-06 PARAmeter     ! SHAKE bonds to
! Hydrogen atoms


! ===============================
! Restart dynamics: Production MD
! ===============================
SET j out/mbco_prep3                       ! Previous job
SET i out/mbco_run                            ! Current job


OPEN UNIT 30 READ FORMatted  NAME @j.res    ! Previous RESTart file
OPEN UNIT 31 WRITe FORMatted  NAME @i.res    ! The RESTart file
OPEN UNIT 32 WRITe UNFOrmatted NAME @i.trj    ! The trajectory file
OPEN UNIT 33 WRITe FORMatted  NAME @i.ene    ! The energy output file


! ********************************************************
! * MODIFY:
! * Run dynamics for 2.0 ps using 1 fs timesteps with no
! * velocity rescaling
! * KEYWORDS to MODIFY: NSTEp, TIMEstep and IEQFrq.
! ********************************************************
```

```
! Production dynamics for 2.0 ps
! ----------------------------
DYNAmics LEAPfrog RESTart -
     NSTEp  ???  TIMEstep  ???  -
     IPRFrq 250  NPRInt   50  FINAlt 300.  -
     IASOrs  0   ISCVel    0  IASVel 1  -
     IEQFrq ???  ICHEcw    0  -
     INBFrq  -1  IHBFrq    0  -
     IUNCrd 32   KUNIt    33  NSAVc 10 NSAVv -1 -
     IUNRea 30   IUNWri   31  ISVFrq 500


! Write the coordinates after Stage 3 of preparation
! -------------------------------------------------
OPEN UNIT 21 WRITe FORM NAME @i.crd
WRITe COOR CARDs UNIT 21
* MbCO: Coordinates after 2.0 ps of production dynamics
*
CLOSe UNIT 21

STOP
```

anal1.tmpl

```
* Lab 6: MbCO Dynamics and Analysis
* Analysis of Average Structure and RMS Fluctuation
*

BOMLev -1

! **********************
! * Modify the file names
! **********************
SET t ???                               ! topology file
SET p ???                               ! parameter file
SET s ???                               ! PSF

! load the topology, parameter and psf files
STREam load_psf.str

! Read the trajectory to analyize
! ------------------------------
SET j out/mbco_run                      ! Previous job
SET i out/anal1                         ! Current job

OPEN UNIT 51 READ UNFOrmatted NAME @j.trj

! calculate average structure and isotropic fluctuation from trajectory
! --------------------------------------------------------------------
COORdinate DYNAmics FIRStunit 51 NUNIt 1


! Write the average structure and isotropic fluctuation onto a
! coordinate file
! (isotropic fluctuation are stored in the last columns of output file)
! --------------------------------------------------------------------

OPEN WRITe UNIT 1 CARD NAME @i_ave.crd
WRITe COORdinate CARD UNIT 1
* Average structure of myoglobin during 2 ps of MD simulation
*

! Use the isotropic fluct. to get atomic type vs rms fluctuation table
! --------------------------------------------------------------------

SCALar WMAIn STORe 1           ! put original fluct. in storage array #1

! Calculate average rms fluctuation of all non-H atoms by segment

SCALar WMAIn AVER BYSEgment SELEct ( .NOT. HYDRogen ) END
SCALar WMAIn SHOW SELEct (RESId 1:5 .AND. SEGId 1MBC .AND. TYPE CA ) END
SCALar WMAIn SHOW SELEct (RESId 1   .AND. SEGId HEME .AND. TYPE C*A) END
```

```
! Calculate average rms fluctuation of all carbon atoms by segment

SCALar WMAIn RECAll 1
SCALar WMAIn AVERage BYSEgment SELEct (TYPE C*) END
SCALar WMAIn SHOW SELEct (RESId 1:5 .AND. SEGId 1MBC .AND. TYPE CA ) END
SCALar WMAIn SHOW SELEct (RESID 1  .AND. SEGId HEME .AND. TYPE C*A) END

! Calculate average rms fluctuation of all CA atoms by segment
SCALar WMAIn RECAll 1
SCALar WMAIn AVERage BYSEgment SELEct (TYPE C*A) END
SCALar WMAIn SHOW SELEct (RESId 1:5 .AND. SEGId 1MBC .AND. TYPE CA ) END
SCALar WMAIn SHOW SELEct (RESID 1  .AND. SEGId HEME .AND. TYPE C*A) END

! Calculate average rms fluctuation of all CB atoms by segment
SCALar WMAIn RECAll 1
SCALar WMAIn AVERage BYSEgment SELEct (TYPE C*B) END
SCALar WMAIn SHOW SELEct (RESId 1:5 .AND. SEGId 1MBC .AND. TYPE CB ) END
SCALar WMAIn SHOW SELEct (RESID 1  .AND. SEGId HEME .AND. TYPE C*B) END

! Calculate average rms fluctuation of all nitrogen atoms by segment
SCALar WMAIn RECAll 1
SCALar WMAIn AVERage BYSEgment SELEct (TYPE N*) END
SCALar WMAIn SHOW SELEct (RESId 1:5 .AND. SEGId 1MBC .AND. TYPE N* ) END
SCALar WMAIn SHOW SELEct (RESID 1  .AND. SEGId HEME .AND. TYPE N* ) END

! Calculate average rms fluctuation of all oxygen atoms by segment
SCALar WMAIn RECAll 1
SCALar WMAIn AVERage BYSEgment SELEct (TYPE O*) END
SCALar WMAIn SHOW SELEct (RESID 1:5 .AND. SEGId 1MBC .AND. TYPE O* ) END
SCALar WMAIn SHOW SELEct (RESID 1  .AND. SEGId HEME .AND. TYPE O* ) END


STOP
```

anal2.tmpl

```
* Lab 6: MbCO Dynamics and Analysis
* Calculation of Internal Coordinate averages and fluctuations
*

BOMLev -1
FASTer on

! **********************
! * Modify the file names
! **********************
SET t ???                          ! topology file
SET p ???                          ! parameter file
SET s ???                          ! PSF

! load the topology, parameter and psf files
STREam load_psf.str

! The PSF file does not carry the IC table, so it is necessary to
! define the internal coordinates which we want to analyze.

IC EDIT

! Back bone near His 64
  DIHEdral 62 N   63 CA   63 C   63 N 0.0
  DIHEdral 63 C   63 N    63 CA  64 C 0.0
  DIHEdral 64 N   64 CA   64 C   65 N 0.0

! Histidine ring motion
  DIHEdral 64 CA  64 CB   64 CG  64 ND1 0.0

! Back bone near His 93
  DIHEdral 91 C   92 N    92 CA  92 C 0.0
  DIHEdral 92 N   92 CA   92 C   93 N 0.0
  DIHEdral 93 C   94 N    94 CA  95 C 0.0

! Tyrosine 103 ring motion
  DIHEdral 103 CA  103 CB   103 CG 103 CD1 0.0

! ***************************************************
! * MODIFY:
! * Define an internal coordinate that characterizes
! * the ring motion of His 93
! ***************************************************
  DIHEdral ???

END

! initialize the IC table
```

A Guide to Biomolecular Simulations   Becker and Karplus

```
IC FILL

SET j out/mbco_run                      ! previous job

! open the trajectory file
OPEN UNIT 51 READ UNFOrmatted NAME @j.trj

! Calculate the averages of the internal coordinates
IC DYNAmics AVERage FIRSt 51 NUNIt 1
IC PRINt
CLOSe UNIT 51

! re-open the trajectory file
OPEN UNIT 51 READ UNFOrmatted NAME @j.trj

! Calculate the fluctuations of the internal coordinates
IC DYNAmics FLUCtuations FIRSt 51 NUNIt 1
IC PRINt

STOP
```

anal3.tmpl

```
* Lab 6: MbCO Dynamics and Analysis
* Correlation Function and its Spectra for CO stretching
*

BOMLev -1
FASTer on

! ***********************
! * Modify the file names
! ***********************
SET t ???                         ! topology file
SET p ???                         ! parameter file
SET s ???                         ! PSF
SET c ???            ! coordinate file

! load the topology, parameter, psf and coordinate files
STREam load_psf.str
STREam load_crd.str

SET j out/mbco_run                ! previous job
SET i out/anal3                          ! current job

! Open the trajectory file
OPEN UNIT 51 READ UNFOrmatted NAME @j.trj

! Open the CORREL module
CORREL MAXTIME 2000 MAXSERIES 1

! Define the property to analyze - the CO stretching motion,
ENTEr CO BOND CO2P 1 C CO2P 1 O GEOMetry

! Read in the trajectory files
TRAJectory FIRSt 51 NUNIt 1 SKIP 10

! Calculate the time series for name CO, defined in the ENTEr command
MANTime CO DAVEr

OPEN UNIT 1 WRITe CARD NAME @j.fluct
WRITe CO UNIT 1 DUMB TIME

! Calculate the correlation function
CORFunction CO CO FFT LTC

OPEN UNIT 2 WRITe CARD NAME @j.corrf
WRITe CORRelation UNIT 2 DUMB TIME

! Calculate the spectrum
SPECTrum SIZE 1500 RAMP SWIT
```

```
OPEN UNIT 3 WRITe CARD NAME @j.spect
WRITe CORRelation UNIT 3 DUMB TIME


STOP
```

Lab 7: Ligand Dynamics in Myoglobin

I. OBJECTIVE

In this exercise you will study the dynamics of a model ligand in the heme pocket of myoglobin. By computing trajectories of unbound ligands you will search for pathways by which a ligand can escape the protein. The number and nature of exit paths will be examined and some of their features analyzed. To simplify the simulation we shall use a spherical model-ligand moving through a rigid protein (instead of a diatomic ligand moving through a flexible protein).

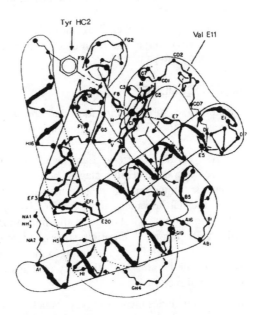

II. BACKGROUND

This lab exercise involves calculating trajectories of ligands in the heme pocket of myoglobin in the spirit of the work of D.A. Case and M. Karplus ("Dynamics of ligand binding to heme proteins", *J. Mol. Biol.* (1979), **132**: 343-368). The aim is to generate many ligand trajectories and then scan for those that escaped from the pocket. The paths taken by the escaping ligands are then examined in terms of the energy profile and structural aspects.

NOTE: You **_must_** read the Case and Karplus paper (*J. Mol. Biol.* (1979), **132**: 343-368) **_before_** tackling this exercise and have it with you when you do!

Some REFERENCES to ligand dynamics in myoglobin:

1.   Kottalam J. and Case D.A.,"Dynamics of ligand escape from the he pocket of myoglobin," JACS (1988),110: 7690-7697.
2.   Brooks, C.L., III, Karplus, M. & Pettitt, B.M. *Proteins: A Theoretical Perspective of Dynamics, Structure and Thermodynamics,* Advances in Chemical Physics 71 (John Wiley & Sons: New York, 1988).
3.   Elber R. and Karplus M., "Enhanced sampling in molecular dynamics: use of time-dependent Hartree approximation for a simulation of carbon monoxide diffusion through myoglobin," *JACS* (1990), **112**: 9161-9175.
4.   Meuwly M., Becker O.M., Stote R., and Karplus M., "NO rebinding to myoglobin: a reactive molecular dynamics study". *Biophys Chem.* (2002), **98**:183–207.

III. PROCEDURE

Create a new sub-directory *lab7* and change to that directory. Copy the content of the *$Lab/lab7* directory to your own directory (**cp -r** *$Lab/lab7/** .).

As you have seen in Labs 5 and 6, molecular dynamics studies involve several successive stages: preparation of the system, production dynamics simulation and finally analysis of the resulting trajectory. To make the lab exercise more productive you will actually perform calculations only in last stage of the simulation; i.e., in the analysis. You will "walk through" the first stages, without actually performing the calculations. Therefore, in addition to the two template files to be used in part C of the lab, the *lab7* directory contains additional files organized in three sub-directories:

lib   includes the topology, parameter, psf and coordinate files.
prep   includes the input files for preparing the system (part A).
runmd includes the input/output to run the MD simulation (part B).

You are required to understand all three stages and answer the questions scattered throughout this lab.

A. PREPARATION OF THE SYSTEM

The first stage involves preparation of the protein-ligand system. You will perform no calculation at this time.

1) The system

The myoglobin coordinates are the X-ray coordinates of T. Takano (*J.Mol.Biol.* (1977), **110**: 537-568 and 569-584) with hydrogens built in by CHARMM. The ligand is represented as a neutral sphere (i.e., monoatomic) with mass and radius appropriate for a dioxygen molecule. The calculations are performed in the *diabatic limit*, in which the protein is kept *rigid* during the molecular dynamics simulation of the ligand's trajectory.

Several measures are taken to reduce (and overcome) the energy barriers that the ligand encounters in the protein:

a)   Ligand radius is decreased.
b)   Ligand is given a high initial kinetic energy.
c)   Multiple ligands are run simultaneously in a single job; the multiple ligands do not interact with each other, only with the protein.

NOTE: The simulation is run at constant total energy (a microcanonical ensemble) and the ligands are moving in the field of a fixed protein. As a result there is no mechanism for energy transfer from the ligands to the protein and the ligands retain their initial total energy throughout the simulation. Namely, the ligands are *hot* thereby having a chance to overcome the exit barriers.

2) Computational steps

The CHARMM input files used to prepare the system described above are located in the *lab7/prep* directory. Take a look at the content of that directory and read through their description below.

a) Build PSF of DxMb (deoxy-myoglobin) – *prep/psfgen.inp*

The PSF of unligated sperm-whale myoglobin is generated from its sequence. A stream file, *helix_name.str*, is used to rename the residues of myoglobin according to the scheme conventionally used for globins. It will be useful in order to compare with Case and Karplus' discussion.

b) Add hydrogen atoms to the DxMb structure – *prep/hbld.inp*

Use HBUIld (two iterations) to generate hydrogen positions. These positions are refined by a short minimization of the new hydrogen coordinates.

c) Adapt topology and parameter files for new ligand entries.

Since this exercise involves model-ligands that were not previously defined, the standard RTF and PARAMETER files had to be changed, in the appropriate places, to include entries for the model ligand. The resulting files were stored in the *lab7/lib* directory under the names: *lib/toph19_lig.inp* and *lib/ param19_lig.inp*.

NOTE: The ligands are defined as atom type LX.

Q1:   Look in the new topology and parameter files and answer the following questions:

i)   What is the mass of the ligand? To what diatomic ligand is it comparable? (look for the MASS keyword in the topology file).

ii)   What is the charge of the ligand? (Look at the RESI LIG definition in the topology file).

iii)   What is the vdW diameter of the spherical model ligand? (In the NBFIX part of the parameter file look for the entries "LX C" and "LX O").

iv)   What is the interaction between two ligands? (In the NBFIX part of the parameter file look for the entry "LX LX").

TIP: The NBFIX part of the parameter file specifies vdW parameters to use for a given pair of atom types. These supersede any parameter that could be generated from standard combination rules. Each line in this section has four elements: the two atomic types involved, E_{min} and R_{min} of the potential.
The total charge is at the end of the RESIdue definition line in the topology file.

d) Add ligand PSF entries and positions – *prep/initial.inp*

Build additional PSF entries for the model ligands, and position all of them at the same point, a few Å away from the heme's iron atom along a vector normal to the plane defined by the pyrolle nitrogens of the heme (atoms NA, NB, NC, ND). A new PSF and a new CRD files are written.

NOTE: The distal side of the heme is +Z in our coordinate system.

Q2: i)   How many ligands are generated in *initial.inp*?

ii)   Each ligand is generated as a separate segment. How are these segments named?

iii)   At what distance from the heme's Fe atom are the ligands placed? (Look at the COOR DUPLicate and COOR TRANslate commands)

iv)   What are the names of the new PSF and CRD files (with ligands)?

```
NEW CHARMM COMMANDS:
   COOR TRANslate    COOR DUPLicate    INCRement
   COOR LSQP         IF ... GOTO ...   LABEl
```

B. PERFORMING MOLECULAR DYNAMICS

Now the molecular dynamics trajectory of the model ligands moving through a fixed protein is calculated. Again, you will perform no actual calculation at this time.

1) Calculate ligand trajectories – *runmd/traj.inp*

Assign a high initial temperature (through initial velocity assignment) such that each ligand has approximately 15 kcal/mol of translational energy, i.e., 15 kcal/mol in the center-of-mass motion (kT ~ 0.6 kcal/mol at 300K). Then perform the dynamics simulation of the ligands in the fixed protein. This *.inp* file also performs some preliminary analysis of the final configuration (records and identifies the ligands that have escaped the heme pocket). We shall return to this part in the next section.

The resulting trajectory, in which the coordinates were saved every 50 time-steps, is stored in the file: *runmd/dxmb_lig.trj* .

Q3:   Look in the *.inp* file and determine what is the length, in picoseconds, of the simulation (identify the time-step and number of integration steps used)? What is the initial temperature used?

C. ANALYSIS OF THE RESULTS

Finally it is time to analyze the results. In this section you will perform the analysis of the resulting ligand trajectories using QUANTA for visual analysis and CHARMM for computational analysis.

1) Escaping vs. non-escaping ligands – *runmd/traj.out*

The *runmd/traj.inp* command file, which was used to perform the dynamics simulation, also includes some initial analysis of the final configuration. This analysis includes:
a) Counting how many ligands have escaped the protein and identifying them.
b) Finding the distances between the non-escaping ligands and the heme's Iron atom.
c) Finding the distances between the escaping ligands and the heme's Iron atom.

The results of this analysis were written to the file *runmd/traj.out*, which is available to you in the *lab7* directory. You are required to extract the relevant results from this file and answer the following questions.

Q4: i)   What is the criterion for "escaping from the protein" that is used in this analysis? How many ligands have escaped and what are their SEGId's? (look for the "DEFIne escape" command)

      ii)   From the *traj.out* file extract the information regarding the distances between the <u>non-escaping</u> ligands and the Iron atom (look for the COOR DIST command). What is the minimal distance? The maximal distance? The average? [If possible **plot** a histogram of the results].

      iii)   The same as question (ii) but for the <u>escaping</u> ligands.

2) Examine ligands global behavior

In this part you will use a graphic visualization program to view the resulting structure and select examples of different exit routes.
The following instructions are for QUANTA, but you can perform similar steps with other visualization programs.

- Create a sub-directory in which to run QUANTA (**mkdir** *quanta*), and move into that directory (**cd** *quanta*).
- Start QUANTA.
- **Import** the initial protein and ligand CHARMM coordinates (RESID from LAST column). Use file *lab7/lib/dxmb_lig.crd*.
- Change the bonding algorithm to Disable Inter-Segment Bonding (**Bond Options** in the **Edit** menu).
- Make a display selection (**Draw** menu) which includes the protein and the ligands that escaped in the trajectory.

Here is an example of a display selection that can be use to view selected trajectories:

> *round 16. cz mb:cd1*
> *excl*
> *type lig*
> *incl*
> *zone L1:1 L3:1 L8:1 L19:1*
> *zone L33:1 L36:1 L40:1 L48:1*
> *zone L53:1 L56:1*

The first part selects a (residue-based) sphere around the CZ atom of the myoglobin residue CD1 = Phenylalanine 43 (located in the heme pocket), then all ligands are excluded from the selection. Finally an explicit selection of the desired ligands is added. [Click on **MORE** to get additional lines to type in the selection]. Note, in the initial coordinate set all ligands overlap each other.

TIP: You can create your own selection files (e.g., *escape.dsd*), which can be read into QUANTA as an external user written display-selection file whenever necessary.

- You may want to make a convenient color selection to increase the clarity of the display. It is suggested that you differentiate at least between heme, protein and ligands. An example of such a selection is:

> *all = col 1*
> *type heme = col 2*
> *type lig = col 3*

- Select **Dynamics Animation** from the **Applications** menu.

- In **Select Trajectories** load the *runmd/dxmb_lig.trj* trajectory file (be sure that you are in the correct sub-directory).
- **Set Up Animation** and download the frames. Re-run the animation using **clock** or **cycle**.
- A nice effect can be made with use of the **Display Trail** option in the dialogue box for **Set Up Animation**.
- Get a feeling for the various escape processes.
- Select 2-3 globally <u>distinct</u> escape paths. Note the ID of the ligands, you will use them in the next section.

3) <u>Examine energetics of selected escape paths</u> - *profile.tmpl*

Now lets study in detail the energetics along those 2-3 escape paths you selected in the previous section. Using the CHARMM template file *profile.tmpl* you will generate, for each of the selected ligands, a *.profile* file. This data file contains the calculated interaction energy between the protein and that specific ligand as a function of time (i.e., along the trajectory).

MODIFY: You have to add some job specifications (e.g., which ligand to follow) and file names to the template files. When necessary, make sure you also specify the name of the sub-directory in which the files are located.

You will have to make a specific *.inp* file (based on the template file) for each of the ligands you want to characterize (2-3 ligands), and then perform the calculation typing (e.g., for ligand L1),

charmm < *profile_L1.inp* > *profile_L1.out* &

A *L1.profile* file will be generated by the job.

Q5:   i)   How is the *profile.tmpl* job terminated? (read the file carefully and distinguish between escaping and non-escaping ligands)

ii)   Discuss the resulting energy profiles (*.profile* files) and identify the times, given in index of data-sets, corresponding to the major energy barriers of the path. Remember you are sampling the trajectory discretely.

iii)   Is the path that the trajectory follows a good example of the reaction coordinate of this process? Discuss.

iv)   What is the largest energy barrier that the ligand has to overcome in its exit? How does this relate to the thermal energy supplied to the ligand?

```
NEW CHARMM COMMANDS:
     COOR MINDistance        TRAJectory
```

4) Details of escape paths - *path.tmpl*

In this section you will further analyze the selected exit paths, and describe the contacts that contribute to the highest barriers for each of them. The template file you are given generates a <u>coordinate file</u> (.*path*), which represents the path of a single ligand, as a pseudo C_α-trace of a molecule. Namely, the coordinate at time frame #1 is defined as residue #1, etc. This enables one to easily visualize the path of a single ligand in QUANTA.

MODIFY: You have to add some job specifications and file names to the template file. When necessary, make sure you also specify the name of the sub-directory in which the files are located.

You will have to make a specific .*inp* file (based on the template file) for each of the ligands you want to characterize (2-3 ligands), and then perform the calculation typing (e.g., for ligand L1),

    charmm < *path_L1.inp* > *path_L1.out* &

A *L1.path* coordinate file will be generated by the job.

Now open QUANTA (or any other visualization program) and analyze the trails you have generated (the .*path* coordinate files).

-     Load the initial DxMb coordinate file (with hydrogens).
-     Load (in addition to the previous structure) the pseudo C_α-trace of your selected ligand exit path (RESId from FIRST column). Select the bonding algorithm **Link** from **Bond Options** under the **Edit** menu (this will provide the necessary connectivity for the trace structure)

Q6: i)   Describe the main features of the trajectory in terms of protein contacts.

-    Use the trace, together with the previously obtained information about the energy profile, to locate the position of the highest barrier.

Q6: ii)   For the (major) barrier(s), list the protein atoms/residues that are within 4.0Å of the ligand.

Two contour plots of the interaction energy of a pseudo-ligand with the DxMb are given below (adapted from Case and Karplus 1979). These plots are in the same coordinate system as our calculation: Figure 1 looks down the z-axis (this is the default orientation), and Figure 2, which is rotated -90 deg about the x-axis from Figure 1, looks down the y-axis (you can achieve this via **Orient View** under **View**).

These plots are useful in describing two of the principal paths found by Case and Karplus.

Q7:   Do your selected trajectories take either of these paths? If so, sketch out the approximate trajectory and indicate the position of the barrier.

COMMENT: Recall that the escape paths and their energetics were obtained using a rigid protein (*diabatic limit*). In general, the next step would be to improve on these results by performing an *adiabatic* minimization of the protein. This is a simple way to relax the protein around one of the barrier configurations, which were found earlier, and remove strains. The adiabatic minimization is performed by picking out of the trajectory file the coordinates that corresponds to the highest barrier along one of the escape paths you studied, and while FIXing the relevant ligand minimizing the protein atoms within a 8 Å radius of that ligand (all other protein atoms being fixed, and the other ligands removed). This process reduces the height of the barrier.

OPTIONAL: You may perform an adiabatic minimization of one of your barriers and earn EXTRA CREDIT. <u>No template files for this one!</u>

Q8:   If you do the adiabatic minimization - To what extent is the barrier reduced? Which atoms have moved the most to accommodate the ligand ?

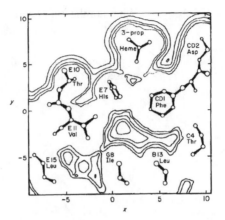

Figure 1:   Contour map of the interaction of a pseudo-ligand with deoxy-myoglobin (DxMb): the (x,y) plane at $z = 3.2$ Å (adapted from Case and Karplus 1979).

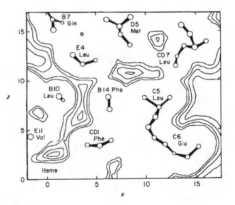

Figure 2:   Contour map of the interaction of a pseudo-ligand with deoxy-myoglobin (DxMb): the (x,z) plane at $y = -0.5$ Å (adapted from Case and Karplus 1979)

IV. WRITE UP

You are expected to answer all the questions posed throughout the lab. In addition please address the following questions:

Q10:   What are the major short-comings of the approach we have taken to examining ligand exit paths?

Q11:   How can these (indeed, have these!) been addressed in subsequent studies?

prep/psfgen.inp

```
*   Lab 7: DxMb - Ligand Dynamics
*   Build PSF for DxMb
*

SET t ../lib/toph19.inp              ! topology
SET p ../lib/param19.inp       ! parameter

SET j ../lib/dxmb                     ! current job name

! load appropriate topology and parameter files
OPEN UNIT 1 READ FORM NAME @t
READ RTF CARD UNIT 1
CLOSe UNIT 1

OPEN UNIT 1 READ FORM NAME @p
READ PARA CARD UNIT 1
CLOSE UNIT 1

! Generate PSF
! DXMb consists of two parts:
!       a. 153 residue polypeptide Mb
!       b. The HEME group

READ SEQUence CARDs
*   Mb (sperm-whale MYOGLOBIN sequence)
*
    153
VAL LEU SER GLU GLY GLU TRP GLN LEU VAL LEU HIS VAL
TRP ALA LYS VAL GLU ALA ASP VAL ALA GLY HIS GLY GLN
ASP ILE LEU ILE ARG LEU PHE LYS SER HIS PRO GLU THR
LEU GLU LYS PHE ASP ARG PHE LYS HIS LEU LYS THR GLU
ALA GLU MET LYS ALA SER GLU ASP LEU LYS LYS HIS GLY
VAL THR VAL LEU THR ALA LEU GLY ALA ILE LEU LYS LYS
LYS GLY HIS HIS GLU ALA GLU LEU LYS PRO LEU ALA GLN
SER HIS ALA THR LYS HIS LYS ILE PRO ILE LYS TYR LEU
GLU PHE ILE SER GLU ALA ILE ILE HIS VAL LEU HIS SER
ARG HIS PRO GLY ASP PHE GLY ALA ASP ALA GLN GLY ALA
MET ASN LYS ALA LEU GLU LEU PHE ARG LYS ASP ILE ALA
ALA LYS TYR LYS GLU LEU GLY TYR GLN GLY

GENERATE MB

READ SEQUence CARDs
*   HEME GROUP
*
    1
```

```
HEME

GENERATE HEME

! introduce the proximal His -> Fe bond
PATCH PHEM MB 93 HEME 1

! renumber the heme as residue 154 of MB
JOIN MB HEME RENUmber

! call the stream file that names the residues according to the globin
scheme.
STREam helix_name.str

! write the PSF
OPEN UNIT 1 CARD WRITe NAME @j.psf
WRITe PSF CARD UNIT 1
*   psf for DxMb
*   rtf @t
*   par @p
*

STOP
```

prep/hbld.inp

```
*   Lab 7: DxMb - Ligand Dynamics
*   Generate Hydrogen Positions
*

BOMLev -1

SET t ../lib/toph19.inp                ! topology
SET p ../lib/param19.in                ! parameter
SET s ../lib/dxmb.psf                  ! psf
SET c ../lib/dxmbx.crd                 ! coordinate from crystal structure

SET j ../lib/dxmb                      ! current job name

! load psf and coordinate files
STREam load_psf.str
STREam load_crd.str


! Add the hydrogen coordinates (two iteration)
HBUIld RDIElectric
HBUIld SELEct (HYDRogen) END

! any coordinates missing ?
DEFIne missing SELEct (.NOT. INITialized) SHOW END

! refine H positions: minimize in a fixed protein
CONStrain FIX SELEct (.NOT. HYDRogen) END
MINImize SD NSTEP 50

! write new coordinates
OPEN UNIT 1 WRITe CARD NAME @j.crd
WRITE COORdinate CARD UNIT 1
* DXMB with hydrogen positions from HBUILD
* 2 interations + 50 steps SD mimimization
*

STOP
```

prep/initial.inp

```
*   Lab 7: DxMb - Ligand Dynamics
*   Set up initial position for LIGands in DxMb
*

BOMLev -1

SET t ../lib/toph19_lig.inp          ! topology
SET p ../lib/param19_lig.inp         ! parameter
SET s ../lib/dxmb.psf                ! psf
SET c ../lib/dxmb.crd                ! coordinates (with hydrogens)

SET j ../lib/dxmb_lig                ! current job name
SET n 100                            ! number of ligands

! loas psf and coordinate files
STREam load_psf.str
STREam load_crd.str


! generate ligand PSF entries
! we will generate them as separate segments for ease of display in quanta
SET 1 1
LABEl gen

   READ SEQUence CARD
* many ligs make light work
*
   1
LIG

GENERate L@1
INCRement 1
IF 1 LE @n GOTO gen

DEFIne ligands SELEct (SEGId L*) END

! generate some coordinates

! define the heme's Fe atom and the 4 pyrolle N as "Nporph"
DEFINE Nporph SELEct (RESName HEME .AND. (TYPE Fe .OR. TYPE N% ) ) SHOW
END

SCALar WMAIN SET 1.0 SELEct (Nporph) END
SCALar WMAIN SET 0.0 SELEct (TYPE Fe) END
SCALar WMAIN SET 0.0 SELEct (ligands) END
```

```
PRINT COOR SELEct (Nporph) END

! get heme normal
COOR LSQP SELEct (Nporph) END WEIGht VERBose

! EQUATION FOR LEAST-SQUARES-PLANE:    -0.0719*X+0.0774*Y-0.9944*Z =       -
0.42


! place the ligands on the Fe atom coordinates (loop over ligands)
SET 1 1
LABEL loop
    COOR DUPLicate SELEct (TYPE Fe) END SELect (SEGID L@1) END
    INCRement 1
IF 1 LE @n GOTO loop


! move the ligands away from the Fe along the normal to the "Nprop" plane

COOR TRANSlate XDIR -0.0719 YDIR 0.0774 ZDIR 0.9944 DISTance 2.5 -
             SELEct (ligands) END


! check with the first ligand
DEFIne setup SELEct (Nporph .OR. (ligands .AND. SEGId L1)) END
COOR   LSQP   SELEct (setup) END WEIGht VERBose
PRINT  COOR   SELEct (setup) END


! write out the PSF and coordinates

OPEN UNIT 1 WRITE FORM NAME @j.psf
WRITE PSF CARD UNIT 1
* psf for dxmb + @n lig
* rtf @t
* par @p
*

OPEN UNIT 1 WRITE FORM NAME @j.crd
WRITE COOR CARD UNIT 1
* DxMb in heme reference frame
* @n LIG ligands placed in arbitrary initial position
* corresponding to an approximately bound geometry
*

STOP
```

runmd/traj.inp

```
*   Lab 7: DxMb - Ligand Dynamics
*   Trajectory dynamics of the model ligands in a FIXEd DxMb
*

BOMLev -1
FASTer ON

SET n 100              ! number of ligands
SET j dxmb_lig         ! job code

SET k 5010.            ! initial temperature for ligands
SET i 314159           ! seed for random number generator (integer)
SET d 2500             ! number of integration timesteps
SET h 0.002            ! time-step in ps

SET t ../lib/toph19_lig.inp    ! topology file
SET p ../lib/param19_lig.inp   ! parameter file
SET s ../lib/dxmb_lig.psf      ! psf
SET c ../lib/dxmb_lig.crd      ! coordinates

STREam ../prep/load_psf.str
STREam ../prep/load_crd.str

DEFIne ligands SELEct (SEGId L*) END

! fix the protein
CONStrain FIX PURGe SELE (.NOT. ligands) END

! open the output trajectory file
OPEN UNIT 77 WRITE UNFOrmatted NAME @j.trj
TITLe
* JOB: @j number of ligs=@n
* Timestep=@h Nstep=@d Initial temp=@k seed=@i
*


! run the MD simulation

DYNAmics LEAPfrog STRT -
        TIMEstep @h NSTEp @d ECHEck 1.0e6 -
        ATOM SHIFt RDIElectric -
        VATOm VSWItch -
        CUTNb 8.5 CTONnb 5.0 CTOFnb 6.0 WMIN 0.0 -
        FIRStt @k TSTRucture @k IEQFrq 0 -
        NPRInt 100 IPRFrq 250 -
        IUNCrd 77 NSAvc 50
```

```
! write out the final coordinates
OPEN UNIT 1 WRITe FORMatted NAME @j.crd
WRITe COOR CARD UNIT 1
* JOB: @j number of ligs=@n
* Timestep=@h Nstep=@d Initial temp=@k seed=@i
* final coordinates
*


! ======================
! SOME   INITIAL ANALYSIS
! ======================

! find out which ligands have escaped

DEFIne escape SELEct ligands .AND. .NOT. -
       (( (SEGid MB .AND. .NOT. HYDRogen) .AROUND. 6.0) .AND. ligands) -
       SHOW END


! distances from iron for non-escaping ligands

COOR DISTance SELEct (ligands .AND. .NOT. escape) END SELEct (TYPE Fe) END


! distances from iron for escaping ligands

COOR DISTance SELEct (escape) END SELEct (TYPE Fe) END

STOP
```

profile.tmpl

```
* Lab 7: DxMb - Ligand Dynamics
* For sample exit trajectories - examine energy profile of the escape path
*

BOMLev -1
FASTer ON


!=======================================================================
! NB: if you want to sample structures less often than saved in the
!     trajectory file - modify the appropriate TRAJectory statement.
!=======================================================================

! ************************************
! * MODIFY:  specify the job parameters
! ************************************
SET j ???                  ! name (and location) of trajectory file
SET l ???                  ! SEGID of ligand of interest (eg. L1)
SET i @l                   ! new job name (used for  creating .profile file)
SET n ???                  ! number of dynamics data-sets in the trajectory
file

! ************************************************************
! * MODIFY the file names (be sure to specify sub-directories)
! ************************************************************
SET t ???                          ! topology file
SET p ???                          ! parameter file
SET s ???                          ! PSF (DxMb + ligands)
SET c ???                          ! initial coordinates (DxMb +
ligands)

! load the psf and coordinate files
STREam prep/load_psf.str
STREam prep/load_crd.str

COOR INIT COMP

! definitions
DEFIne ligands SELEct (SEGId L*) END
DEFIne target  SELEct (ligands .and. SEGID @l) END

! Fix all atoms except for the one to analyze
CONStrain FIX PURGe SELEct (.NOT. target) END

! open the trajectory file
OPEN UNIT 77 READ UNFOrmatted NAME @j
TRAJectory IREAd 77 NREAd 1
```

```
! open the .profile output file
OPEN UNIT 11 WRITe FORMatted NAME @i.profile
WRITE TITLE UNIT 11
* Trajectory: @j Ligand: @l
* time     distance   energy
* index    from Fe    (kcal/mol)
*


! =================================================================
! loop over the data-sets in the trajectory and calculate Fe-Ligand
! distance and the energy.
! =================================================================

SET    1 1              ! loop index
LABEl  loop

   TRAJectory READ      ! read a data-set from the trajectory file

   COOR MINDISTance SELEct (target) END SELEct (TYPE Fe) END

   ENERgy -
        ATOM SHIFt RDIElectric VATOm VSWItch -
        CUTNb 7.5 CTONnb 4.0 CTOFnb 5.0 WMIN 0.0

! write the results
   WRITe TITLe UNIT 11
* @l ?MIND ?ENER
*


! check if ligand has escaped
   DEFIne escape SELEct target .AND. .NOT. -
      (( (SEGid MB .AND. .NOT. HYDROgen) .AROUND. 8.0) .AND. ligands)   SHOW
END

   IF ?NSEL GT 0.0 GOTO escape

   INCRement 1

IF 1 LE @n GOTO loop


! finish

LABEl escape

CLOSe UNIT 11

STOP
```

A Guide to Biomolecular Simulations   Becker and Karplus

path.tmpl

```
* Lab 7: DxMb - Ligand Dynamics
* create a coordinate file to describe ligand's path
*

! In this job a coordinate file ( .path ) is created. These coordinates
! are a dummy CA trace with consecutive residues corresponding to the
! trajectory of a selected ligand. It is just a convenient way to view the
! results in QUANTA as the path of the trajectory will be traced out and
! individual points can be geometrically interrogated.

BOMLev -1
FASTer ON


!=========================================================================
! NB: if you want to sample structures less often than saved in the
!     trajectory file - modify the appropriate TRAJectory statement.
!=========================================================================

! ************************************
! * MODIFY:  specify the job parameters
! ************************************
SET j ???              ! name (and location) of trajectory file
SET l ???              ! SEGID of ligand of interest (eg. L1)
SET i @l               ! new job name (used for  creating .path file)
SET n ???              ! number of dynamics data-sets in the trajectory
file

! **************************************************************
! * MODIFY the file names (be sure to specify sub-directories)
! **************************************************************
SET t ???                        ! topology file
SET p ???                        ! parameter file
SET s ???                        ! PSF (DxMb + ligands)
SET c ???                        ! initial coordinates (DxMb +
ligands)

! load the psf and coordinate files
STREam prep/load_psf.str
STREam prep/load_crd.str

COOR INIT COMP

! definitions
DEFIne ligands SELEct (SEGId L*) END
DEFIne target  SELEct (ligands .and. SEGID @l) END
```

```
! constrain the protein
CONStrain FIX PURGe SELE (.NOT. ligands) END

! { get intial position}
COOR COPY COMP SELEct (target) END
COOR DUPLicate COMP SELEct (target) END SELEct (IRES 1 .AND. TYPE CA) END

! open the trajectory file
OPEN UNIT 77 READ UNFOrmatted NAME @j
TRAJectory IREAd 77 NREAd 1

! ================================================================
! loop over the data-sets in the trajectory and copy the ligand's
! coordinate to that of a CA at the residue number that equals the index
! of the data-set.
! ================================================================

SET    1    2                    ! loop index
LABEl loop
    TRAJectory READ
    COOR COPY COMP SELEct (target) END
    COOR DUPLicate COMP SELEct (target) END SELEct (IRES @1 .AND. TYPE CA)
END

! check for escape from the protein
    DEFIne escape SELEct target .AND. .NOT. -
    (( (SEGid MB .AND. .NOT. HYDRogen) .AROUND. 8.0) .AND. ligands)   SHOW
END

    IF ?NSEL GT 0.0 GOTO escape

    INCRement 1
IF 1 LE @n GOTO loop


! finish

LABEL escape

COOR COPY
DEFIne used SELEct (TYPE CA .AND. INIT) END

! clean up the atoms not "used"
DELEte ATOM SELE (.NOT. used) END

RENAMe RESName LIG SELEct (ALL) END

! write out the ligand's path as a coordinate file
OPEN UNIT 1 WRITe FORMatted NAME @i.path
WRIT    1E COOR CARD UNIT 1
```

```
* trajectory: @j LIGAND @l
* ligand's path as pseudo CA trace
*

STOP
```

I. OBJECTIVE

In this lab you will perform normal mode analysis of molecules and study the resulting frequencies and normal coordinates. These will be discussed in terms of degeneracy, in/out of plane motion, contribution of internal coordinates and more.

   Since normal mode analysis of a real protein is computationally intensive and can be very complicated, this lab will deal with a simpler system: the two related molecules benzene and phenol. Through these molecules you will become familiar with the normal mode analysis procedure and see the role of symmetry and degeneracy on the vibrational spectrum.

II. BACKGROUND

Internal motions of many-particle systems (such as proteins) can be described as a superposition of their 'fundamental' motions, which are called *normal modes* (or vibrational modes). The goal of Normal Mode Analysis is to calculate these fundamental modes and explore them.

   In general, any molecule of N atoms can be characterized by 3N degrees-of-freedom (e.g., 3 Cartesian coordinates for each atom). When transformed to a center-of-mass system, one finds that of these 3N degrees-of-freedom, 3 characterize global translation of the center-of-mass, 3 characterize rigid global rotation of the whole molecule, and 3N-6 degrees-of-freedom characterize internal vibrations. These 3N-6 degrees-of-freedom are the normal modes of the system.

Normal Mode Analysis

Normal Mode Analysis follows the notion that although the Cartesian coordinate system is very useful it may not be the *natural* coordinate system for the molecule. There may be a different coordinate system in which the description of the molecular motion will be far simpler. This other coordinate system is the system of *normal modes*. Thus, what 'normal mode analysis' actually does is to transform the coordinate system from the regular Cartesian set to a *normal* set, which is more *natural* to the molecule.

155

HARMONIC APPROXI- MATION

In practice, Normal Mode Analysis employs a very strong approximation known as the "harmonic" approximation. The approximation assumes that the shape of the potential near the minima is quadratic in all coordinates. Using this assumption it is easy to show that a coordinate transformation from the original coordinate set $\{q_i\}$ to a new normal set $\{Q_i\}$ is possible, and that the potential in this new coordinate set is separable.

NORMAL MODES

$$V(q_1, ..., q_N) ==. \Sigma_i V(Q_i)$$

The meaning of separability is that in the new coordinate systems $\{Q_i\}$ all the normal modes are independent (and harmonic). These can be used as building blocks for more complex motions.

Technically the transformation is performed by simultaneously diagonalizing the system's Hessian matrix; i.e., the matrix of second derivatives of the potential with respect to the original coordinates ($H_{ij} = \partial^2 V(q_1,...,q_N)/\partial q_i \partial q_j$), and the system's mass matrix. The eigenvalues of the resulting diagonal matrix give the frequencies of the normal modes (i.e., the spectrum) and the eigenvectors describe the displacements associated with the normal modes.

BASIC REFERENCES

- Goldstein, H., *Classical Mechanics* (Addison-Welsley 1980), Chapter 6.
- Levitt, M., Sander C. and Stern, P.S. "Protein normal-mode dynamics: trypsin inhibitor, crambin, ribonuclease and lysozyme", *J. Mol. Biol.* **181** (1985), 423-447.

Analysis of results

Analysis of the resulting eigenvalues and eigenvectors can tell much about the nature of the molecular vibrations and their spectroscopic properties.

Among the various properties that can be studied are:
1) Vibrational Spectrum - the collection of all normal frequencies. This spectrum is comparable to experimental vibrational spectra taken with IR spectroscopy, Raman spectroscopy or Neutron diffraction. Differences between the calculated spectrum and the experimental one may reflect selection rules or inaccuracies in the potential surface and unharmonic effect.
2) Degeneracy - two or more modes that have identical frequencies are said to be degenerate. Degeneracy is a mark of high symmetry in the molecule. It is removed when symmetry is broken.

3) Geometric characteristics of the motion - studying the eigenvectors that describe the modes one learns about fundamental motions in the molecule. In small molecules, such as those studied in this lab, we see in-plane versus out-of-plane vibrations, modes that primarily involve stretching of the C-H bonds versus global puckering or twisting modes and more. In larger systems (e.g., proteins) global motions such as hinge bending or helix twisting can be identified.

4) IR activity - a change in the molecular dipole moment along a normal mode is a strong indication that this mode is IR active (i.e., it absorbs IR radiation).

Limitations of Normal Mode Analysis

Normal Mode Analysis is a very powerful technique, which reveals much about the molecule. Nevertheless, it does suffer from several limitations:

1) The harmonic approximation, which is the cornerstone of this method, is valid only for small vibrational amplitudes. Large vibrational amplitudes often involve anharmonic effects, which cannot be accounted for by this method. Fortunately, some protein-scale delocalized concerted motions are of small vibrational amplitudes and can be viewed as low frequency normal modes. Delocalized concerted motions involve small vibrational amplitudes of hinged domains and can be approximated by low frequency normal modes.

2) The analysis has to be performed around a minimum on the potential energy surface. If not at a minimum the analysis will yield negative frequencies, which do not have the regular meaning of vibrational frequencies. Therefore, to get a valid result the system <u>must</u> be strictly minimized before calculating the Hessian matrix to be diagonalized.

3) Carrying out the analysis requires matrix diagonalization, which usually scales as N^3 with the size of the system. As a consequence, the size of systems that can be studied with this technique is limited. New algorithms have been developed to overcome this computational limitation (see the reference to Simonson and Perahia 1992).

4) Like all other methods studied in this lab, Normal Mode Analysis cannot be more accurate than the underlying empirical potential energy function.

REFERENCES to some Normal Mode Analysis studies of proteins:

1. Brooks, B., and Karplus, M., "Harmonic dynamics of proteins: normal modes and fluctuations in bovine pancreatic trypsin inhibitor", *Proc. Natl. Acad. Sci. USA* (1983), **80**: 6571-6575.

2. Go, N., Noguti, T., and Nishikawa, T., "Dynamics of a small globular protein in terms of low-frequency vibrational modes", *Proc. Natl. Acad. Sci. USA* (1983), **80**: 3696-3700.

3. Brooks, B. and Karplus, M. "Normal modes for specific motions of macromolecules: application to the hinge-bending mode of lysozyme", *Proc. Natl. Acad. Sci. USA* (1985), **82**: 4995-4999.

4. Simonson, T. and Perahia, D. "Normal modes of symmetric protein assemblies: application to the tobacco mosaic virus protein disk", *Biophys. J.* (1992), **61**: 410-427.

III. PROCEDURE

Create a new sub-directory *lab8* and change to that directory. Copy the content of the *$Lab/lab8* directory to your own directory (**cp -r** *$Lab/lab8/* .*).

This directory includes the topology, parameter and coordinate (ideal geometry) files for benzene and phenol as well as several *.tmpl* template files and *.str* stream files (the usual *load_psf.str* and *load_crd.str*). You will also find there a sub-directory named *quanta*, which already contains two MSF files (*benzene.msf, phenol.msf*). To facilitate the exercise create two new sub-directories (**mkdir**), one named *benzene* and the other named *phenol*, and copy all the files into both sub-directories (e.g., **cp** * *benzene/*.).

You will perform the same exercise for both molecules, first benzene and then phenol. It includes the following steps:

1) Generate PSF and IC table.
2) Perform normal mode analysis of unminimized molecule.
3) Minimize the molecule.
4) Perform normal mode analysis of the minimized molecule.
5) Analyze the modes.
6) Write trajectory files for several modes.
7) Use a visualization program to view the normal modes graphically.

To run the jobs use:

charmm < *command_ file.inp* > *output_ file.out* &

A. BENZENE

Move to the *benzene* sub-directory you have created (**cd** *benzene*).

1) Generate PSF and IC table from the coordinate file - *gen.tmpl*

This job generates a PSF file using the supplied coordinate file. It also saves the IC-table for future use. Make sure you know which file is which.

MODIFY: You have to add job specifications and several file names to the template file.

2) Perform initial normal mode analysis - *norm1.tmpl*

Perform a basic normal mode analysis of the benzene molecule. Since, at this stage, we are interested only in the frequencies, the only command necessary in CHARMM's VIBRan command block is DIAGonalize.

COMMENT: The DIAGonalize command performs the diagonalization of the Hessian matrix, and lists as an output all the frequencies.

MODIFY: Add some job specifications and file names to the template file.

Q1:   Look at the frequencies obtained. How many frequencies are not appropriate for normal modes (i.e., negative frequencies)? What is the remedy? (For the purpose of this question any frequency with an absolute value smaller than 10^{-3} should be considered as equal to zero).

3) Minimize the molecule - *mini.tmpl*

Although the input coordinate file reflects a perfect geometry, the CHARMM parameters do not exactly match it. For a small molecule like benzene (or phenol) the simple Steepest Descent minimization should do. The *mini.tmpl* template file performs 500 SD minimization steps.

MODIFY: Add some job specifications and file names to the template file.

To see the results of the minimization, perform the following command on the *output_file.out*:

grep -i 'mini>' *output_file.out* | **more**

The column before last gives the GRMS. It is advisable to get down to GRMS < 0.001 (press the space-bar to see the next screen).

Q2:   i)   How many SD minimization steps were performed?
        ii)   What was the final GRMS?
        iii)   Offer your opinion on the outcome of the minimization and on the number of minimization steps used (too much? too little?).

4) Normal mode analysis of the minimized molecule - *norm2.tmpl*

Use the coordinates obtained from the minimization as an input to the normal mode analysis. Since this time we want a detailed analysis, be sure to include the various analysis facilities in the VIBRan block (see below). Especially important are the PRINt NORM command and the PED command.

 MODIFY: In addition to the job specifications and file names you usually add to the template file, you also have to specify the keywords in the PRINT NORM command. Refer to the following TIPS and use the CHARMM documentation.

 COMMENT: Benzene has 36 degrees of freedom (12 atoms), which correspond to 36 normal modes. Modes number 1 through 6 correspond to the six global translation-rotation degrees-of-freedom. They do not represent vibrations and their computed 'frequencies' are *zero* (to the numerical accuracy of the calculation).

 TIPS: 1)   Be sure to use the coordinates of the minimized structure.
         2)   The PRINt NORM command gives information about the coordinates of the normal mode (the eigenvector).
               Several keywords govern the output, some useful ones, which you should include, are:
               VECTors - to get the normal coordinates in Cartesian coordinates.
               DIPoles - changes in dipole value during the mode.
               STATistics - some vector statistics.

> INTDer - internal coordinate derivatives; i.e., the change in IC associated with a mode. Large values indicate an Internal Coordinate that changes significantly during the motion.

3) The PED (= Potential Energy Distribution) command computes what would be the changes in energy if the system moved along a normal mode. These changes are specified for each internal coordinate (bond, angle, dihedral and improper dihedral) and the output includes information on any term for which the expected fluctuation is greater than the tolerance (TOL). The command works on a selection of modes, one mode at a time.

4) Both PRINt NORM and PED commands require an *amplitude specification*, which can be either in terms of temperature (TEMP) or of mass weighted RMS (MRMS). Make sure that the *amplitude specification* is large enough so that the motion is visible (e.g., MRMS 1.0).

5) Analysis of the modes

As you work through the following tasks and questions write the information to a table similar to the one below. In the first three columns (frequency, degeneracy & in/out of plane) fill in for *all* the modes, modes with a degeneracy should appear only once. In the remaining columns fill in the details only for modes 7 through 16 and modes 31 through 36.

| mode Nº. | Frequency | Degeneracy | in/out of plane motion | IR activity | IC character | | | type |
					Bonds	Angles	Dihd. & Impr.	
7/8	676.61	2	in	-	20.4	79.6	-	CC, CCC
9/10	757.35	2	out	-	-	-	100.0	

a) Frequencies and Degeneracy

First study the frequencies that were obtained. Identify the translation and rotation "modes". Regarding the remaining 3N-6 modes list all frequencies and their degeneracy in the above table (Columns 1-3). The criterion for degeneracy should be a frequency difference on the order of 0.01 cm^{-1}. Now answer the following questions:

Q3:   i)   How many modes are doubly degenerate?
           ii) How many are not degenerate?
           iii) Are there any modes that are triply degenerate?

b) In-plane/out-of-plane motions

Look at the eigenvector coordinates (normal coordinate in Cartesian) and in the table describing changes in Internal Coordinates (especially the dihedral angles). Identify the modes that are IN-plane and those that are OUT-of-plane and write them in the appropriate column.

TIP: Out-of-plane modes are characterized by significant changes in the z-component of the eigenvector. Since the molecule is planar it must also involve changes in the dihedral angles ϕ.

Q4:   i)   How many in/out modes are there?
           ii) How many of these are singly/doubly degenerated?
           iii) Are there modes that are both in-plane and out-of-plane?

c) IR activity

Look at the *Total Dipole Derivative* for modes 7-16 and 31-36. If the value is larger than 0.001 it means that the mode induces a significant change in the dipole moment, which in turn makes it IR active. Indicate this activity in the "IR activity" column.

TIP: IR activity occurs when light interacts with the molecule. A study of these interactions reveals that IR absorbance can occur only if the molecule has a changing dipole moment (selection rule). Therefore, only modes with a non-zero dipole-derivative are IR active.

Q5:   i)   How many of these eigenvectors have a changing dipole?
           ii) How many different modes (i.e., allowing for degeneracy) have a changing dipole?
           iii) Are these in- or out-of plane modes?

d) IC character

Look at the information obtained from the PED command regarding the potential energy distribution for modes 7-16 and 31-36. Fill the three columns under "IC character" with the "total bond contribution", "total angle contribution" and the sum of the "total dihed." and "total impr.". Also, for the in-plane modes use the more detailed IC information to decide what type(s) of bonds or angles contribute most to the mode (use the notation CC, CH, CCC, CCH). Add this in the "type" column.

e) Pure stretch or bend

Using the information collected above (and listed in the table), look at all the *in-plane modes* and identify:

Q6:   i)   Which of these are pure stretching modes (CC or CH)?
         ii)   Which are pure bending modes (CCC or CCH)?

f) Out-of-plane modes

Now look at the *out-of-plane* modes found within the range of modes studied (7-16 and 31-36). Using the information in the eigenvector coordinates and in the IC derivative table identify the nature of *each* of these normal-coordinates.

Q7:   i)   Draw a cartoon with arrows indicating the direction and magnitude (schematically) of each C and H atom. Identify each mode as Puckering/Twist etc. Use the notation "syn" if the H moves in the same direction as the C atom it is bound to, and "anti" if they move in opposite directions.

         For example: Mode no. 9 is a "puckering mode (syn)":

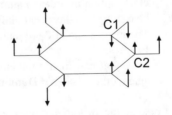

ii)   Why is the "Ring Inversion" mode (mode number 15)
         associated with a changing dipole while the other out-of-
         plane modes are not?

6) Write trajectory files for several modes – *traj.tmpl*

By generating artificial dynamic trajectories along the normal coordinates
we can look at the normal mode motion on the graphics display. We
suggest that you generate such *.trj* files for modes 9 to 22, but any other
range is just as good. The resulting trajectories should be written directly
to the *lab8/quanta* sub-directory. The trajectory files should be named
quanta/{molecule}_{mode_number}.trj, e.g. the trajectory of mode no. 7
of benzene should be named *quanta/benz_7.trj* .

MODIFY: Add some job specifications and file names to the template
               file. Use the minimized structure and choose the range of
               modes you want to analyze.

7) View the normal modes graphically

- Go to the *quanta* sub-directory (**cd**) and start QUANTA.
- Open the appropriate MSF file, that is supplied with the directory (use
  the dials to rotate the molecule out of the {xy} plane for better
  viewing).
- Choose **Dynamics Animation** from the **Applications** menu.
- From the *Dynamics Animation* pallet choose **Set up Animation**. Select
  the option **Display Dipole for Whole Molecule** and set speed to 50.
  Click OK.
- Choose **Select Trajectories**, then **Initialize dynamics files**, click OK,
  enter the trajectory file name; Exit.
- Choose **Create Animation**. After it completes one cycle, select the
  clock option in order to animate the continuous motion (the speed can
  be changed on the speed dial).
- To finish a specific animation de-select clock and **Delete Dynamics
  Frames**.
- Now you can load another trajectory file using **Select Trajectories**
  again, otherwise **Exit Dynamics Animation**.

Look at the different modes and verify the information you extracted
previously from the CHARMM output.

B. PHENOL

Move to the *phenol* sub-directory you have created (**cd** *phenol*).

Repeat all the above steps (1)-(7) for the phenol molecule. Answer *all* the questions and tasks along the way. This time answer in detail about modes 7 through 16 and modes 34 through 39. You are invited to skip section (5.f), i.e., do not draw the cartoons.

TIP: You may find it easier to group the IR activity to Weak/Moderate/Strong according to the change in dipole (smaller than 1.0; between 1.0 and 1.7; larger than 1.7).

IV. WRITE-UP

The fact that there is no degeneracy in phenol arises from the lower symmetry of the molecule (compared to the high symmetry of benzene). It can be said that the addition of the OH group "perturbs" the benzene structure and as a result the degenerated modes are split into two distinct modes.

Answer the following questions:

Q8:   Write side by side the frequencies associated with the <u>pure</u> CH stretches in benzene and the <u>pure</u> CH/OH stretches in phenol (6 normal-coordinates). Draw a line from each phenol frequency to its "predecessor" in the benzene (if possible). Use other properties, such as IR activity, to help in this "mapping". The splitting in this case is easy to identify.

Q9:   What happened to the totally symmetric CH stretch in benzene (mode 36) upon addition of the OH in phenol. Is mode 39 of phenol a direct "descendent" of mode 36 in benzene? What is the character of phenol's mode 39?

Q10:   Why are almost all the modes in phenol IR active, while only relatively few modes in benzene were IR active?

Q11:   Identify the character of the two lowest modes in phenol (modes 7,8). Use all the tools available, i.e., eigenvector coordinates, IC derivatives, PED results, and graphics animation.

gen.tmpl

```
* Lab 8: Normal Mode Analysis of small molecules
* Generate psf and ic-table of a molecule from its coordinate file
*

! ****************************
! MODIFY the job specification
! ****************************
SET n ???        ! name of the generated molecule (benz or phen)

! ********************
! MODIFY the file names
! ********************
SET t ???              ! topology
SET p ???              ! parameter
SET c ???              ! coordinates (ideal geometry)

! { setup the system }
! ====================
! { topology }
OPEN UNIT 1 CARD READ NAME @t
READ RTF CARD UNIT 1
CLOSe UNIT 1

! { parameters }
OPEN UNIT 1 CARD READ NAME @p
READ PARAmeter CARD UNIT 1
CLOSe UNIT 1

! { read the sequence from the coordinate file }
OPEN UNIT 1 CARD READ NAME @c
READ SEQUence COORdinate UNIT 1
CLOSe UNIT 1

! { generate the molecule }
GENErate @n SETUp WARNing

! { write psf file }
OPEN UNIT 2 CARD WRITe NAME @n.psf
WRITe PSF CARD UNIT 2
* PSF file of @n (all hyrdogen)
* TOP: @t  , PAR: @p
*
CLOSe UNIT 2

! { write ic file }
OPEN UNIT 3 CARD WRITe NAME @n.ic
WRITe IC CARD UNIT 3
* IC file of @n (all hyrdogen)
* TOP: @t  , PAR: @p
*

STOP
```

norm1.tmpl

```
* Lab 8: Normal Mode Analysis of small molecules
* Performing basic Normal Mode Analysis (on unminimized structure)
*

! *****************************
! MODIFY the job specifications
! *****************************
SET n ???      ! name of the  molecule (benz or phen)
SET m ???      ! number of degrees of freedom (3*N)

! ********************
! MODIFY the file names
! ********************
SET t ???           ! topology
SET p ???           ! parameter
SET s ???           ! psf
SET c ???           ! coordinates (ideal geometry)

! load the psf and coordinate files
STREam../load_psf.str
STREam../load_crd.str


! { initialize the energy and forces arrays}
FASTer OFF
NBONd CUTNb 999 CTOFnb 998 CTONnb 997 CDIElectric  ! no cutoffs


ENERgy


! { perform normal mode analysis }
! ================================

VIBRan
        DIAGonalize

END
```

mini.tmpl

```
* Lab 8: Normal Mode Analysis of small molecules
* Minimization using 500 SD steps (no harmonic restrains)
*

! *****************************
! MODIFY the job specifications
! *****************************
SET n ???        ! name of the  molecule (benz or phen)
SET j ???        ! name of OUTPUT file with minimized coordinates (no
siffix)

! *********************
! MODIFY the file names
! *********************
SET t ???                 ! topology
SET p ???                 ! parameter
SET s ???                 ! psf
SET c ???                 ! coordinates (ideal geometry)


! load the psf and coordinate files
STREam../load_psf.str
STREam../load_crd.str


! set non-bonded specifications (no cutoffs)
NBONd CUTNb 999 CTOFnb 998 CTONnb 997 CDIElectric


! { minimize the structure }
! ==========================
MINImize SD NSTEp 500 IHBFrq 0 INBFrq -1


! { print the minimized coordinates }
! ==================================
OPEN UNIT 1 CARD WRITe NAME @j.crd
WRITe COORdinate CARD UNIT 1
*  Coordinate of @n after 500 SD minimization steps
*

STOP
```

norm2.tmpl

```
* Lab 8: Normal Mode Analysis of small molecules
* Performing detailed Normal Mode Analysis (on minimized structure)
*

! *****************************
! MODIFY the job specifications
! *****************************
SET n ???      ! name of the  molecule (benz or phen)
SET m ???      ! number of degrees of freedom (3*N)

! ********************
! MODIFY the file names
! ********************
SET t ???              ! topology
SET p ???              ! parameter
SET s ???              ! psf
SET c ???              ! coordinates (minimized structure)
SET i ???              ! IC file


! load the psf and coordinate files
STREam../load_psf.str
STREam../load_crd.str


! { read the IC table and fill it according to the crd file }
OPEN UNIT 1 READ CARD NAME @i
READ IC CARD UNIT 1
CLOSe UNIT 1

IC FILL

! { initialize the energy and forces arrays}
FASTer OFF
NBONd CUTNb 999 CTOFnb 998 CTONnb 997 CDIElectric  ! no cutoffs

ENERgy

! { perform detailed normal mode analysis }
! ==========================================

! ****************************************************************
! MODIFY: Add several options and "magnitude-specifications" to the
!         PRINT NORM command. Add "magnitude-specifications" to the
!         PED command. Refer to the TIPS in your LAB MANUAL and the
!         CHARMM MAUAL for directions.
! ****************************************************************

VIBRan
        DIAGonalize

        PRINt NORM ???

        PED MODE 7 THRU @m TOL 0.01 ???
END
```

traj.tmpl

```
* Lab 8: Normal Mode Analysis of small molecules
* Creating Trajectory files for specific modes
*

! ======================================================================
! Writes trajectories for MODE @f THRU @l . The trajctories are
! written sequentially to fortran units starting from 21, and they
! are named ../quanta/{molecule}.trj{no. of mode}, e.g. the trajectory
! of mode no. 7 of benzene is written to: "../quanta/benz.trj7".
! ======================================================================

! *****************************
! MODIFY the job specifications
! *****************************
SET n ???      ! name of the molecule (benz or phen)
SET f ???      ! the first mode (to make a trajectory from)
SET l ???      ! the last  mode (to make a trajectory from)


! ********************
! MODIFY the file names
! ********************
SET t ???            ! topology
SET p ???            ! parameter
SET s ???            ! psf
SET c ???            ! coordinates   (minimized structure)


! load the psf and coordinate files
STREam../load_psf.str
STREam../load_crd.str


! set non-bonded specifications (no cutoffs)
NBONd CUTNb 999 CTOFnb 998 CTONnb 997 CDIElectric


! initialize the energy and forces arrays
FASTer OFF
ENERgy


! { open a trajectory files for each desired modes }
! =================================================
SET k 21       ! first unit number
SET i @f       ! current mode index

LABEl loop
    OPEN WRITe UNIT @k FILE NAME ../quanta/@n_@i.trj
    INCRement i
    INCRement k
IF i LE @l GOTO loop


! { start normal mode analysis and write trajectories )
```

```
! ========================================================

VIBRan
        DIAGonalize

        WRITe TRAJectory UNIT 21 MODE @f THRU @l SEQUential MRMS 0.5
END

STOP
```

I. OBJECTIVE

In this lab you will learn how to calculate thermodynamic properties of biomolecules, in particular free energy differences. The methodology of such free energy calculations, which use molecular dynamics simulations, will be discussed and the quality and stability of the results will be analyzed. For convenience the calculations in this lab we be performed on a simplified but representative molecular model system.

II. BACKGROUND

Thermodynamic properties, such as free energy, entropy and enthalpy, of complex systems describe much of the behavior of these systems. Thus, it is an important task, for experimentalists and theoreticians alike, to understand and determine thermodynamic quantities for any system of interest. As most thermodynamic quantities are "state functions" the focus is usually on their *difference* between two well-defined states rather than their absolute value. Examples include the free energy difference between two protein conformations (e.g., native versus denatured), differences in the free energy of unfolding between wild type and mutant proteins (i.e., stability of mutant compared to wild type), free energy of solvation, and the free energy difference between bound and dissociated protein ligand complexes in solution.

The foundation for these thermodynamic properties is the statistical-mechanical canonical partition function Z of the system, which is based on the individual state energies E_k

PARTITION FUNCTION

$$Z = \sum_k e^{-\beta E_k}$$

with $\beta = 1/k_B T$, where k_B is the Boltzmann constant and T is the absolute temperature; the summation is over all possible states k of the system. The free energy of the system and all other thermodynamic properties are derived from the partition function, e.g., the Helmholtz free energy A is defined as

HELMHOLTZ FREE ENERGY

$$A = -(1/\beta)\ln Z$$

Once the partition function Z has been determined, all macroscopic properties, including free energies, are easily derived from it. Unfortunately, evaluating the partition function for large systems is a *very* hard task. Consequently, methods have been devised to directly calculate free energy differences (ΔA) rather then derive the total free energy (A) from the partition function (Z).

A powerful method for solving this problem is thermodynamic perturbation theory [e.g., Kollman 1993]. The main idea of this theory is that the potential energy function can be partitioned in a convenient way.

THERMO-DYNAMIC PERTURBATION

One can write

$$V(\lambda) = V_0 + \lambda V_\lambda$$

where V_0 represents the potential for a suitably defined 'reference system' and V_λ is the perturbation relating the reference system to the system of interest (such as a point mutation in a protein). In this expression V_λ is independent of λ (depends only on the system coordinates) and λ is the perturbation parameter, $\lambda = 1$ for the fully perturbed system and $\lambda = 0$ for the reference state. However, unlike classical perturbation theory, in protein studies the unperturbed system is not necessarily easier to treat than the perturbed system (for example, the perturbation could be the energetic contribution of a point mutation in a large protein or binding of a small ligand). Thus, in large molecular systems we use a slightly different partition of the energy

$$V(\lambda) = V_0 + V_{AB}(\lambda) = V_0 + (1-\lambda) V_A + \lambda V_B$$

where V_0 is the part that is identical for both states and

$$V_{AB}(\lambda) = (1-\lambda) V_A + \lambda V_B$$

includes the perturbation which is λ-dependent. V_A includes interactions unique to the initial state and V_B includes those interactions that are unique to the final state, and V_A and V_B are invisible to each other, i.e., non-interacting. For example, when studying the free energy difference due to a point mutation in a protein, say alanine to glycine, V_A represents alanine specific interactions, V_B represents glycine specific interactions and V_0 represents the rest of the system that is not affected by the mutation.

Exponential Formula

One way to calculate conformational free energy differences, known as the **Exponential Formula** [Zwanzig 1954], stems from the basic definition of the partition function and the free energy (see above). This formula can be written as

EXPONENTIAL
FORMULA
$$\Delta A_{BA} = A_B - A_A = -(1/\beta) \ln < \exp [-\beta (\Delta V_{BA})] >_A$$

To evaluate this expression we define intermediate states λ_i and calculate the free energy change between two such states in a similar way

$$\Delta A_{\lambda i} = A_{\lambda i+1} - A_{\lambda i} =$$
$$-(1/\beta) \ln < \exp [-\beta (V_{AB}(\lambda_{i+1}) - V_{AB}(\lambda_i))] >_{\lambda i}$$

where $\lambda_{i+1} = \lambda_1 + \Delta\lambda$. The ensemble average, depicted by $< ... >_{\lambda i}$, is obtained by performing a Molecular Dynamics simulation of the system at that value of λ. In practice, to make the computation more efficient, you will perform the simulation at a λ value between λ_{i+1} and λ_i (e.g., if λ_{i+1} = 0.1 and λ_i = 0.0, the simulation will be performed at λ = 0.05). The total free energy difference between state A and state B will be the sum of these intermediate contributions,

$$\Delta A_{BA} = \sum_{i=1}^{N_i} \Delta A_{\lambda_i}$$

where N_i is the number of intermediate steps, and usually $\lambda_i = 0$ and λ_{Ni} = 1 - $\Delta\lambda$.

Thermodynamic Integration

A slightly different, though closely related formula, is obtained by differentiating the free energy A with respect to λ and then integrating back [Kirkwood 1935]. After differentiating and using standard thermodynamic relations we get

THERMO-
DYNAMIC
INTEGRATION
$$\frac{\partial(\beta A_{AB}(\lambda))}{\partial\lambda} = \beta \left\langle \frac{\partial V_{AB}(\lambda)}{\partial\lambda} \right\rangle_\lambda = \beta \langle \Delta V \rangle_\lambda$$

where, as above, $V_{AB}(\lambda) = (1 - \lambda) V_A + \lambda V_B$ and $\Delta V = V_B - V_A$ (independent of λ) and the brackets $< ... >_\lambda$ imply a canonical ensemble average obtained through Molecular Dynamics simulation using the value λ. At each different λ value the average of the perturbation potential is determined from an equilibrium dynamic trajectory of the system.

Integrating back this expression gives the overall free energy difference

$$\Delta A_{AB} = A_B - A_A = \int_0^{\lambda} \langle \Delta V \rangle_{\lambda'} d\lambda'$$

This method is called **Thermodynamic Integration**. In practice, the integration is replaced by a summation

$$\Delta A_{BA} = \sum_{i=1} \langle \Delta V \rangle_{\lambda_i} \Delta\lambda$$

A more detailed derivation of the above formula can be found in references 1 through 3 below, and in Brooks, M. Karplus and B. M. Pettitt, *Proteins: A Theoretical Perspective of Dynamics, Structure and Thermodynamics*, Advances in Chemical Physics LXXI (John Wiley, New York 1988), Ch. IV.

Analysis and Pitfalls

In practical application of thermodynamic calculations there are many pitfalls. To start with, a problem may arise when the perturbation is 'large' (even a mutation of glycine to alanine in water or a protein environment is 'large'). In these cases the desired averages converge very slowly. Care must be taken to carry the MD runs to long enough time to enable the system to equilibrate at each new λ value and to thoroughly sample the ensemble. Only the later part of each such trajectory is used in the analysis.

HYSTERESIS

One indicator for the quality of the thermodynamic calculation is to calculate the free energy once from A to B, and then repeat the calculation going from B to A. Since free energy is a "state function" the two results should be identical (but with opposite signs). In practice there is always a difference between the two values. The accumulated errors are a measure of quality of the simulation protocol (also known as the *hysteresis*). It should be noted, that the actual hysteresis is sometimes masked by fluctuations and should be looked at only after sufficient averaging is performed on both forward and backwards calculations.

In addition, the pathway of intermediate steps has to be carefully chosen. First of all, as both methods discussed above use a discrete sum instead of an integral the more intermediates the better the results. This however has to be balanced against computational efficiency, which requires as few steps as possible. In practice these two demands are balanced and typically a value of $\Delta\lambda = 0.1$ is used.

One possible advantage of computational thermodynamic studies over experimental studies is that on the computer we can breakdown the free energy change into different *contributions* of groups of atoms (e.g., amino acid residues) or of types of interactions (van der Waals, electrostatic and bonding interactions) [Boresch et al. 1994]. However, this kind of breakdown is problematic as it turns out that the resulting analysis is *path dependent*. Namely, while the resulting value of the free energy (which is a state function) is *independent* of any specific path the different contributions that make up this value depend on the way we do the calculation. The individual contributions along a path A → C are different from the contributions along a path A → B → C, although the total free energy is identical. Taking this into account, one can still use this component analysis to analyze the properties of different feasible paths!

PATH DEPENDENT

The Model System

The model system chosen for this lab is a pair of double ionized molecules, putrescine (1,4-diamino-butane or DAB) and 5-amino-valeric acid (AVA), and the related singly charged compound amino-butane (AB) that lacks one of the terminal carboxy or amino groups (see Figure 1). This system was suggested by Brady and Sharp [ref. 5]. Since the focus is on exemplifying the method, the extended atom representation (CHARMM-19) is used and the two charged groups (NH_3^+ and COO^-) are defined as united atoms (with appropriate mass and charge (see the topology file *data/topo.inp*). In addition bond lengths and bond angles are also held fixed (at 1.54 Å and 109.4^0 respectively). As a result, the bonded interactions in this model system are defined solely by three torsion angles around the C1-C2, C2-C3 and C3-C4 bonds. The dynamics of the system is defined by these interactions and by the non-bonded interactions (or lack thereof in AB) between the two terminal functional groups. This model system is in vacuum.

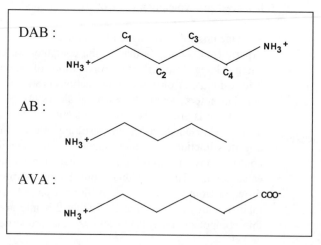

Figure 1: The model system molecules: 1,4-diamino-butane (putrescine, DAB), Amino-valeric acid (AVA) and amino-butane (AB).

REFERENCES:

1. Kollman, P. "Free energy calculations: applications to chemical and biochemical phenomena", *Chem. Rev.* **93** (1993), 2395-2417.
2. Zwanzig, R.W. "High-temperature equation of state by a perturbation method. I. Nonpolar Gases", *J. Chem. Phys.* **22** (1954), 1420-1426.
3. Kirkwood, J.G. "Statistical mechanics of fluid mixtures", J. Chem. *Phys.* **3** (1935), 300-313.
4. Boresch, S. Archontis G. and Karplus, M. "Free energy simulations: the meaning of the individual contributions from a component analysis", *Proteins* **20** (1994), 25-33.
5. Brady G.P. and Sharp, K.A. "Decomposition of interaction free energies in proteins and other complex systems", *J. Mol. Biol* **254** (1995), 77-85.

III. PROCEDURE

Create a new sub-directory *lab9* and change to that directory. Copy the content of the *$Lab/lab9* directory to your new directory (**cp -r** *$Lab/lab9/* *** .). Type **ls** to see the content of the directory. In addition to the input files (*.inp*) there is a *data* sub-directory that contains the parameter and topology files (*para.inp* and *topo.inp*). The main *lib9* directory also includes two 'awk' script files (*pert.awk* and *pert1.awk*) that will help you in the analysis.

A. Free Energy Calculation

Using the supplied files, you will now perform a thermodynamic calculation of the free energy change associated with the mutation DAB → AVA. The calculation is performed using the PERT module in CHARMM ("PERT" stands for "perturbation"). In this implementation of the above-described methods *two* PSFs (Protein Structure Files) are setup side by side, each consisting of a Single Topology. One PSF represents the initial state V_A while the other represents the final state V_B. The dynamic simulation uses both PSFs to compute the free energy at each λ value, with λ ranging from 0.0 to 1.0.

1)   *dab2ava.inp*

Read through the input file and identify: (i) where the bonds and angles are being constrained (using SHAKE); (ii) where the second PSF is being setup; and (iii) where the dynamic integration is defined.

Q1:   What is the difference between the two PSFs set-up in the simulation?

NOTE: 1) To sample the dynamics in a *canonical* ensemble we use Langevin dynamics (i.e., stochastic dynamics) at 300 K and not the *microcanonical* Newtonian dynamics used so far. For more details on Langevin dynamics see: C.L. Brooks, M. Karplus and B.M. Pettitt, *Proteins: A Theoretical Perspective of Dynamics, Structure and Thermo-dynamics*, Advances in Chemical Physics 71 (John Wiley, New York 1988), Ch. 4.

2) Note the keyword ISEED @r in the DYNAmics command. This keyword determines the 'seed' for the random number generator used in the integration. To exactly reproduce the same calculation the same 'seed' must be used. The actual value will be put in the variable @r from the command line (see below).

2) *pert.inp*

This file defines the $\Delta\lambda$ steps of the thermodynamic integration and how to perform them. The lines look like:

```
lstart 0.0 lstop 0.1 pstart 15000 pstop 20000 pwind lambda
0.05
```

This kind of line should be read as follows: "For the interval $\lambda = 0.0$ to $\lambda = 0.1$, run 20,000 integration steps but accumulate data only between step 15,000 and 20,000; use the value $\lambda = 0.05$ for running the ensemble dynamics and for calculating the free energy of this step."

Q2:   What is the time step used (see *dab2ava.inp*)? What is the total length of the simulation? In each step, what is the length of time for which data is accumulated?

3) Run the job

Create a sub-directory *scratch* to which files that are generated during the calculation will be directed (**mkdir** *scratch*).

run the job in the following way

      charmm r=<number> < *dab2ava.inp* > *dab2ava.out*

The <number> stands for any number that you want (e.g., the number 1). This number will be the 'seed' for the random number generator used in the integration.

NOTE: The execution of this job can take a few minutes (how many minutes depends on your computer).

4) Analyze the results

The job you just ran calculated the free energy using both methods described above: the Exponential Formula and the Thermodynamic Integration.

Look at the output file.

To analyze the results it is advised first to extract them from the large output file. For this purpose you are supplied with a prepared **awk** script: *pert.awk*. What this script does is to extract from the large CHARMM output file only the lines which are necessary for the analysis.

Run this script in the following way:

awk **-f** *pert.awk dab2ava.out > dab2ava.dat*

In the output file (*dab2ava.dat*) lines starting with the string 'PERTURBATION> TP ' give the <u>incremental</u> free energy change associated with that step (EFORWARD) and the <u>accumulated</u> free energy (EPRTOT) using the <u>Exponential Formula</u>. Lines containing the string 'PERTURBATION> TI ' give same information calculated using the <u>Thermodynamic Integration</u> method.

Q3: Draw a graph of the incremental free energy as a function of λ (use the LAMBDA value) for both methods. Comment on the way the free energy changes along the path. Compare the two methods.

Q4: Draw a similar graph for the accumulated free energy. Compare the two methods. Estimate the error as (TP - TI)/TP in percentages. Where TP is the final "exponential formula" result and TI is the final "thermo-dynamic integration" result.

B. The stability of the Free Energy Calculation

STABILITY

In this part of the lab you will analyze the stability of the above free energy calculation. First, you will compare the 'forward' DAB → AVA thermodynamic integration with the 'backward' AVA → DAB calculation. Then you will check the effect of using different (random) initial conditions for the molecular dynamics portion of the calculation.

NOTE: Deviations of 1-2 kcal/mol are to be expected in such simulations.

BACKWARD
INTEGRATION

I. 'Backward' integration

To prepare the necessary CHARMM input file make a copy of the *dab2ava.inp* file (e.g., **cp** *dab2ava.inp dab2ava_b.inp*).

Using your favorite editor introduce the following changes into the new copy:

1) In the "ENERgy ... LAMBda 0.0" line change the initial λ value from 0.0 to 1.0.
2) In the "OPEN UNIT 20 ..." line change the name of the file that defines the λ-steps from *pert.inp* to *pert_b.inp*.

Look in the *pert_b.inp* file. As you can see, the steps are now defined 'backwards', starting from λ = 1 and going down to λ = 0.

Run the new input file as described above, using the same 'seed' number. Use the same **awk** script to extract the results from the output.

Q5:   Compare the end result of the 'forward' integration and the 'backward' integration as calculated with both methods. Comment on the differences and estimate the errors. Is the difference you see a measure of the hysteresis? What, in your mind, can be the origin of the differences?

II. Effect of 'seed' for random number generator

Repeat the 'forward' integration *four* more times, each time using a <u>different</u> 'seed' for the random number generator (e.g. seeds $r = 1, 3, 5, 7, 9$). Extract the results.

NOTE: If you are using a slow workstation consider reducing the number of repeated 'forward' integrations to expedite the lab.

Q6:   Construct a table comparing the end results obtained by the two methods for all five 'forward' simulations. Calculate the average and standard deviation of the results. Comment. Also, for each job calculate the error between the TP and TI results (as defined in **Q4**).

NOTE: After answering **Q6** go back and check your answer to the final portions of **Q5**.

COMMENT: The quality of the calculation you encounter in this lab is due to the small size of the system that is susceptible to fluctuations. In larger, and more realistic systems, the quality of the results is usually better.

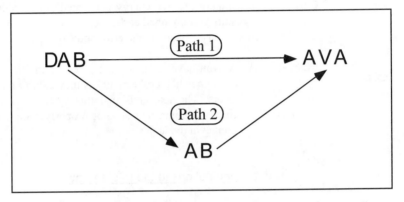

Figure 2:   Two possible paths connecting DAB and AVA: a direct path (path 1) and a path via the intermediate AB (path 2).

C. Comparing Two Paths

Free energy is a state function and therefore is path independent. Namely, the overall free energy associated with the DAB → AVA transition should be independent on whether it is calculated directly, or with an intermediate. Figure 2 depicts two such possible paths: the direct one ('path 1') and a path via the intermediate AB ('path 2'). The sum of the free energy change along the two legs of 'path 2' should equal that of 'path 1'.

1) DAB → AB

Calculate the first leg of 'path 2'. Look in the input file *dab2ab.inp*. Identify the two PSFs that are being setup in the simulation.

Run the job as before, and **awk** out the results.

2) AB → AVA

In a similar way calculate the second leg of 'path 2' using the input file
ab2ava.inp.

Q7:   Look in the *ab2ava.inp* file. Is the integration really
                 performed as AB → AVA? How is it actually done? Does
                 it matter?

Q8:   Compare the free energy calculated along 'path 2' with the
                 results you obtained earlier for 'path 1' (use both calculation
                 methods). Estimate the error and Comment.

COMMENT: As you have seen, so far the results in our model system
                 are somewhat 'problematic' as they exhibit relatively large
                 errors. The reason is that this system has several local
                 energy minima that are not properly sampled during the
                 Langevin dynamics run.

D. Individual contributions to the free energy

It was mentioned above that, in principle, the free energy changes
calculated on the computer can be broken down into *contributions* of
different types of interactions (when using the Thermodynamic
Integration method). There is a debate on the meaning and usefulness of
this breakdown (which is path dependent).

The output files obtained so far also include the breakdown of the
free energy into interaction terms. Due to the nature of our simplistic
model system there are only three types of interactions present: dihedral
torsion, electrostatic and van der Waals.

To help you extract the desired contributions to the free energy, a
special **awk** script has been prepared (*pert1.awk*). This script extracts the
desired information from any of your *output-file*. To use it type

    awk -f *pert1.awk* <u>*output-file*</u>

Extract the different interaction contributions to the free energy as
calculated along 'path 1' and as calculated along the two segments of 'path
2'.

Q9:   Construct a table comparing the total free energy and its components along the two paths (sum up path 2). Comment on similarities and/or differences.

IV. WRITE UP

Answer all the above questions. In addition:

Q10:   EXTRA CREDIT: As you have seen, the results you obtained have a large error associated with them. Suggest ways to improve these calculations. You are encouraged to refer to: Hodel A., Rice L.M., Simonson T., Fox R.O., and Brunger A.T., "Proline cis-trans isomerization in staphylococcal nuclease: multi-substrate free energy perturbation calculations", *Protein Science* **4** (1995), 636-654.

dab2ava.inp

```
* Lab 9: Thermodynamic calculations
* Free energy calculation for DAB -> AVA
*

! Read topology and parameter files

OPEN UNIT 1 READ FORMmatted NAME data/topo.inp
READ RTF CARD UNIT 1
CLOSe UNIT 1

OPEN UNIT 1 READ FORMatted NAME data/para.inp
READ PARAmeter CARD UNIT 1
CLOSe UNIT 1

! Generate the DAB molecule

READ SEQUence CARD
* di-amino butane
*
1
dab

GENErate dab SETUp NOANGL NODIHE

! build coordinates based on the parameters

IC PARAmeter
PRINt IC
IC SEED 1 n1 1 c2 1 c3
IC BUILd

PRINT COORdinates

ENERgy

! Hold bonds and angles fixed at their "parameter" value

SHAKe BOND ANGLe PARAmeter TOLI 1.e-8 MXIT 250

ENERgy INBFrq 1

! Select atoms to be included in the calculations (all atoms(

BOMLev -4
PERTurbation SELEct ( ALL ) END
BOMLev  0

! Setting up the second PSF:
! Define the changes from the first (DAB) PSF to the second (AVA) PSF:
! i.e., change the sixth group in DAB (NH3+) to COO.-

SCALar TYPE   SET    2 SELEct ( ATOM dab 1 n6 ) END
SCALar CHARge SET -1.0 SELEct ( ATOM dab 1 n6 ) END

ENERgy INBFrq 1 LAMBda 0.0
```

```
OPEN UNIT 11 WRITe FORMatted NAME scratch/junk.rst
OPEN UNIT 20 READ  FORMatted NAME pert.inp

! Setup and run the Thermodynamic Integration loop using Langevine
! Dynamics.

SCALar FBETA SET 100 SELEct ( ALL ) END

DYNAmics LEAPfrog LANGevine START -
         NSTEp 110000  TIMEst 0.001  NPRInt 1000   IPRFrq 5000 -
         FIRStt   240. FINAlt  300. -
         PUNIt     20  ISEEd   @r  -
         IUNCrd    -1  IUNRead 10   IUNWrite 11 -
         ILBFrq    10  RBUFfer 0.0  TBATh   300. INBFrq 0

STOP
```

pert.inp

```
* basic "punit" file for WINDOWING
*

lstart 0.0 lstop 0.1 pstart  15000 pstop  20000 pwind lambda 0.05
lstart 0.1 lstop 0.2 pstart  25000 pstop  30000 pwind lambda 0.15
lstart 0.2 lstop 0.3 pstart  35000 pstop  40000 pwind lambda 0.25
lstart 0.3 lstop 0.4 pstart  45000 pstop  50000 pwind lambda 0.35
lstart 0.4 lstop 0.5 pstart  55000 pstop  60000 pwind lambda 0.45
lstart 0.5 lstop 0.6 pstart  65000 pstop  70000 pwind lambda 0.55
lstart 0.6 lstop 0.7 pstart  75000 pstop  80000 pwind lambda 0.65
lstart 0.7 lstop 0.8 pstart  85000 pstop  90000 pwind lambda 0.75
lstart 0.8 lstop 0.9 pstart  95000 pstop 100000 pwind lambda 0.85
lstart 0.9 lstop 1.0 pstart 105000 pstop 110000 pwind lambda 0.95

end
```

pert_b.inp

```
* basic "punit" file for WINDOWING
* backward direction
*

lstart 1.0 lstop 0.9 pstart  15000 pstop  20000 pwind lambda 0.95
lstart 0.9 lstop 0.8 pstart  25000 pstop  30000 pwind lambda 0.85
lstart 0.8 lstop 0.7 pstart  35000 pstop  40000 pwind lambda 0.75
lstart 0.7 lstop 0.6 pstart  45000 pstop  50000 pwind lambda 0.65
lstart 0.6 lstop 0.5 pstart  55000 pstop  60000 pwind lambda 0.55
lstart 0.5 lstop 0.4 pstart  65000 pstop  70000 pwind lambda 0.45
lstart 0.4 lstop 0.3 pstart  75000 pstop  80000 pwind lambda 0.35
lstart 0.3 lstop 0.2 pstart  85000 pstop  90000 pwind lambda 0.25
lstart 0.2 lstop 0.1 pstart  95000 pstop 100000 pwind lambda 0.15
lstart 0.1 lstop 0.0 pstart 105000 pstop 110000 pwind lambda 0.05

end
```

dab2ab.inp

```
* Lab 9: Thermodynamic calculations
* Free energy calculation for DAB -> AB
*

! Read topology and parameter files

OPEN UNIT 1 READ FORMmatted NAME data/topo.inp
READ RTF CARD UNIT 1
CLOSe UNIT 1

OPEN UNIT 1 READ FORMatted NAME data/para.inp
READ PARAmeter CARD UNIT 1
CLOSe UNIT 1

! Generate the DAB molecule

READ SEQUence CARD
* di-amino butane
*
1
dab
GENErate dab SETUp NOANGL NODIHE

! build coordinates based on the parameters

IC PARAmeter
PRINt IC
IC SEED 1 n1 1 c2 1 c3
IC BUILd

PRINT COORdinates

ENERgy

! Hold bonds and angles fixed at their "parameter" value

SHAKe BOND ANGLe PARAmeter TOLI 1.e-8 MXIT 250

ENERgy INBFrq 1

! Select atoms to be included in the calculations (all atoms(

BOMLev -4
PERTurbation SELEct ( ALL ) END
BOMLev  0

! Setting up the second PSF:
! Define the changes from the first (DAB) PSF to the second (AB) PSA:
```

```
! i.e., change the fifth group from CH2 to CH3 (=type 4) and the
! sixth group is effectively removed (defined as a dummy atom,=type 5).

SCALar TYPE    SET     4 SELEct ( ATOM dab 1 c5 ) END
SCALar TYPE    SET     5 SELEct ( ATOM dab 1 n6 ) END
SCALar CHARge  SET   0.0 SELEct ( ATOM dab 1 n6 ) END

ENERgy INBFrq 1 LAMBda 0.0

OPEN UNIT 11 WRITe FORMatted NAME scratch/junk.rst
OPEN UNIT 20 READ  FORMatted NAME pert.inp

! setup and run the Thermodynamic Integration loop using Langevine
! Dynamics.

SCALar FBETA SET 100 SELEct ( ALL ) END

DYNAmics LEAPfrog LANGevine START -
        NSTEp 110000  TIMEst .001  NPRInt 1000   IPRFrq 5000 -
        FIRStt   240. FINAlt  300. -
        PUNIt     20  ISEEd    @r -
        IUNCrd    -1  IUNRead  10  IUNWrite 11  -
        ILBFrq    10  RBUFfer 0.0  TBATh    300. INBFrq 0

STOP
```

ab2ava.inp

```
* Lab 9: Thermodynamic calculations
* Free energy calculation for AB -> AVA
*

! Read topology and parameter files

OPEN UNIT 1 READ FORMmatted NAME data/topo.inp
READ RTF CARD UNIT 1
CLOSe UNIT 1

OPEN UNIT 1 READ FORMatted NAME data/para.inp
READ PARAmeter CARD UNIT 1
CLOSe UNIT 1

! Generate the AVA molecule

READ SEQUence CARD
* amino valeric acid
*
1
ava
GENErate ava SETUp NOANGL NODIHE

! build coordinates based on the parameters

IC PARAmeter
PRINt IC
IC SEED 1 n1 1 c2 1 c3
IC BUILd

PRINT COORdinates

ENERgy

! Hold bonds and angles fixed at their "parameter" value

SHAKe BOND ANGLe PARAmeter TOLI 1.e-8 MXIT 250

ENERgy

! Select atoms to be included in the calculations (all atoms(

BOMLev -4
PERTurbation SELEct ( ALL ) END
BOMLev  0

! Setting up the second PSF:
! Define the changes from the first (AVA) PSF to the second (AB) PSA:
! i.e., change the fifth group from CH2 to CH3 (=type 4) and the
! sixth group is effectively removed (defined as a dummy atom,=type 6.(

SCALar TYPE    SET    4 SELEct ) ATOM ava 1 c5 ) END
SCALar TYPE    SET    6 SELEct ( ATOM ava 1 c6 ) END
SCALar CHARge SET  0.0 SELEct ( ATOM ava 1 c6 ) END

ENERgy INBFrq 1 LAMBda 1.0
```

```
OPEN UNIT 11 WRITe FORMatted NAME scratch/junk.rst
OPEN UNIT 20 READ  FORMatted NAME pert_b.inp

! setup and run the Thermodynamic Integration loop using Langevine
! Dynamics.

SCALar FBETA SET 100 SELEct ( ALL ) END

DYNAmics LEAPfrog LANGevine START -
        NSTEp 110000  TIMEst .001  NPRInt 1000   IPRFrq 5000 -
        FIRStt  240. FINAlt  300. -
        PUNIt    20  ISEEd    @r  -
        IUNCrd   -1  IUNRead  10  IUNWrite 11 -
        ILBFrq   10  RBUFfer 0.0  TBATh   300. INBFrq 0

STOP
```

Lab 10:   Minimum Energy Paths and Transition States

I. OBJECTIVE

This lab introduces 'minimum energy paths' (also known as 'reaction paths') in molecular systems. A 'minimum energy path' is the least energy path connecting two minima on the molecular potential energy surface. The potential energy along this path is called the 'energy profile' and its highest points correspond to saddle-points, also known as 'transition states'. Knowledge of minimum energy paths, energy profiles and transition states contribute to understanding both the dynamics and the kinetics of the molecule.

This lab includes two exercises: (1) Calculating the minimum energy path connecting the 'chair' and 'boat' conformations in cyclohexane, and (2) calculating a more complex energy path, with several saddle-points, in alanine-tetra-peptide (IAN).

II. BACKGROUND

REACTION PATH

MINIMUM ENERGY PATH

A 'reaction path' or 'minimum energy path' describes the path that the molecule is most likely to take during a transition from one state to another. In terms of the potential energy function it is defined as a set of points (conformations) connecting the initial conformation (e.g., the 'chair' conformation of cyclohexane) with the final conformation (e.g., the 'boat' conformation of cyclohexane). Since the potential energy is a multi-dimensional surface, the transition state is defined as a saddle point in the full multidimensionality (see Figure 1).

In general, one should distinguish between two types of 'reaction paths'. The first is associated with changes in the electronic state (like reactions that involve bond breakage or formation) and the other is associated with conformational changes (such as the 'chair' to 'boat' transition in cyclohexane). Although in both cases the reaction path has the same meaning, in the 'reactive' changes it involves crossing between two different electronic states while in 'conformational' changes the path is defined on a single potential surface. In this lab we will deal only with 'conformational' changes.

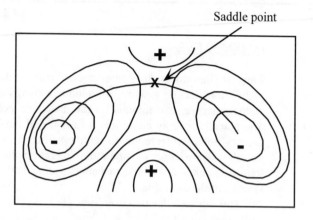

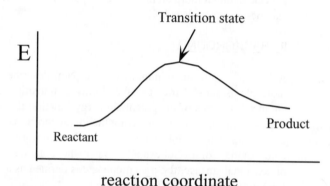

reaction coordinate

Figure 1: A schematic representation of a reaction path and its potential
        energy profile.

Determining 'minimum energy paths' and transition states is a difficult
task. Methods for determining 'minimum energy paths' (or 'adiabatic path')
fall into two categories: 'global' methods and 'focal' methods. Global
methods treat the entire path as a 'string' of points in conformation space
(e.g., Elber and Karplus 1987; Ulitsky and Elber 1990). These points are
simultaneously minimized under constraints that keep them connected.
Focal methods, on the other hand, modify a single conformation until it
becomes the saddle point (e.g., Fischer and Karplus 1992). The path itself is
then obtained by a steepest descent procedure from that saddle point.

Conjugated   The algorithm that will be used in this lab is the Conjugated Peak
Peak   Refinement (CPR) of Fischer and Karplus (1992), which is implemented in
Refinement

TRAVEL   CHARMM as the *TRAVEL* module. This method is geared to finding the series of saddle points along a minimum energy path. The minimum input consists of the reactant and product structures and the output is a string of points in conformation space that follow the minimum energy path.

REFERENCES

1. Fischer S. and Karplus, M. "Conjugate peak refinement: an algorithm for finding reaction paths and accurate transition states in systems with many degrees of freedom", *Chem. Phys. Lett.* **194** (1992), 252-261.
2. Elber R. and Karplus, M. "A method for determining reaction paths in large molecules: application to myoglobin", *Chem. Phys. Lett.* **139** (1987), 375-380.
3. A. Ulitsky and R. Elber, "A new technique to calculate steepest descent paths in flexible polyatomic systems". *J. Chem. Phys.* **92** (1990), 1510-1511.

III. PROCEDURE

Create a new sub-directory *lab10* and change to that directory. Copy the content of the *$Lab/lab10* directory to your new directory (**cp -r** *$Lab/lab10/** .). Typing **ls** you will see that this directory contains two sub-directories: *CYCLOHEXANE* and *IAN*.

To run the jobs use:

> **charmm** < *command_file.inp* > *output_file.out* &

A. Cyclohexane

In this exercise you will find the minimum energy path associated with the 'chair' to 'boat' conformational transition in cyclohexane.

First move to the *CYCLOHEXANE* directory (**cd** *CYCLOHEXANE*). Type **ls** to see the content of the directory; it includes the topology and parameter files (*cycl.rtf, cycl.par*) and two coordinate files (*CHAIR.CRD* and *TWBOAT.CRD*). You will also find in the directory a special **awk** script file that will help you in the analysis (*anal_cyclo.awk*)

The CHARMM input file *cyclo.inp* is designed to calculate the minimum energy path of the above transition and identify the saddle point along it. Read through the file and understand its general scheme, then run the job as described above. It may take a few minutes to run.

Identify the two additional files that are created by this job: a trajectory file that contains the path (*cyclo.trj*) and the coordinates of the saddle point (*saddle.crd*).

COMMENTS: 1) The topology file has the molecule defined as 'cycl' (look in the topology file to see that).

2) You can read more about the specific algorithms employed (CPR, SDP and SCM) in the CHARMM documentation and in Fischer and Karplus (1992).

Analysis

1) To help you extract the relevant data out of the output file you are supplied with a special **awk** script file that will do it for you. To run it type

   awk -f *anal_cyclo.awk output_file > your_file*

where *output_file* is the output file that was just generated and *your_file* is any file name that you choose.

2) Read through the output of the analysis script. Identify in it the following items: IC tables of three conformations (chair, twisted-boat and the saddle-point), information about the saddle point energy and RMS gradient and two tables describing the minimum energy paths after the first and the second refinement (the second is much more detailed).

The minimum energy path tables look like this:

```
Analyzing the reaction-coordinate :
====================================
N   Idx  Length rms(xN-x1) Energy rms(Grad) LINMN Angle GrdProj
--- --  ------ ---------- ------ --------- ----- ----- -------
1   1 0.000E+0 0.000E+0 4.296207E+0 9.710E-4  0  ***** 0.000E+0
2  22 7.297E-3 7.297E-3 4.323017E+0 1.377E-1  0 72.708 8.775E-2
3  21 3.907E-2 3.465E-2 4.693517E+0 1.112E+0  0 12.367 1.058E+0
... ....  .....
```

where Length is the position of that point along the path (in Å), rms(xN-x1) is the rms distance of that point from the first structure (here the 'chair' conformation), and rms(Grad) the gradient rms at that point.

Q1:   Write down the energy and RMS gradient of the saddle-point after the first and second refinements. Comment on the quality of the convergence after the second refinement.

Q2:   Draw the energy profile of this transition using both the unrefined and the refined paths. Use the Length values for the 'reaction coordinate' axis. Comment on the necessity of the refinement. Compare this energy profile to the results you obtained in Lab 2 (especially questions 1 and 2).

Q3:   Draw a similar energy profile, now using the RMS distance of each point to the 'chair' conformation (the rms(xN-x1) column). Use only the refined data. What is peculiar about this graph (hint: near the 'twisted boat' conformation). Explain how this peculiarity can come about, and why this RMS measure is not a good choice for the 'reaction coordinate' axis.

3) Use QUANTA (or any other visualization program) to compare the three coordinate sets (*CHAIR.CRD*, *TWBOAT.CRD* and *saddle.crd*).

- **Import** the three coordinate sets into QUANTA. You will need to specify a dictionary. Select "generic organic and heteroatom functional groups".
- Use **Molecular Similarity** (under **Applications**) to overlay the three structures (**Match Atoms** → **Select All Atom Pairs**, and disable the Display of the matches; then **Rigid Body Fit to Target**). Record the RMS distances from the *Textport*.
- From the *Molecular Similarity* pallet choose **Flash Molecules**, and see how the saddle-point conformation is indeed a intermediate conformation between the chair and the twisted-boat conformations. Color each confirmation differently for a better view. Now use **Molecular Animation** to animate the reaction path (*cyclo.trj*).

B. IAN

Now you will tackle a more complicated energy path between two conformations of the molecule alanine tetra-peptide (the official name of this molecule is isobutyryl-(ala)$_3$-NH-methyl, and it is shown in Figure 2). This path will be calculated in two steps: initial evaluation of the full path, and then additional refinement of different sections of the path.

1) Move to the *IAN* directory (**cd** *IAN*). Type **ls** to see the content of the directory. You will find there the topology and parameter files (*ian.rtf*, *ian.par*), two coordinate files (*conf1.crd* and *conf2.crd*), and a STREAM file (*toppar.str*) for loading the topology and parameter files. You will use the *initial.inp*, *final.tmpl* and *anal_ian.awk* files to execute the exercise.

2) *initial.inp* -

Use this input file to calculate the first calculation of the path connecting
the two conformations *conf1* and *conf2*. First read through the input file
and answer the following question.

Q4:   What is the electrostatic model used in this calculation? Is a
                  cutoff applied to the non-bonded terms (think!)?

Now run the job as described above (depending on your machine, it may
take between a few seconds to a few minutes).

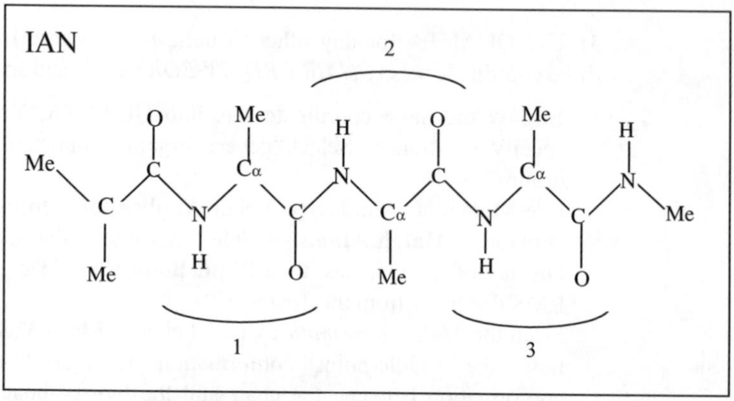

Figure 2:   The isobutyryl-(ala)3-NH-methyl molecule (IAN), also
              known as alanine tetra-peptide.

3) Use the supplied **awk** script (*anal_ian.awk*) to extract the path
information from the large output file,

   awk -f *anal_ian.awk output_file > your_file*

Q5:   Draw the energy profile along this path. Use the Length
                  value for the 'reaction coordinate' axis. Draw the Energy
                  axis *relative* to the lowest energy in the system. How many
                  saddle-points can you identify? how many intermediate
                  conformations? (do not consider small 'humps' as distinctive
                  saddle-points of intermediates). Record the 'point-number'

(N) along the path associated with minimal energies (including the two edge conformations).

4) *final.tmpl* -

Now we want to refine the path. Because there are several saddle-point along this path, the final refinement has to be performed in sections. Each section containing a single saddle-point and stretches from the minimum on one side of the saddle-point to the next minimum on its other side.

✳ Copy the template file to an *.inp* file and enter the required MODIFICATIONS. These include the two points defining the section of the path to be refined (the 'point-number' N), and the name of the trajectory file which holds the initial-path (look in *initial.inp* for its name).

Run this job and then use the **awk** script *anal_ian.awk* to extract the refined path (for that section).

Repeat this process for all sections of the initial path (each section containing one saddle-point).

Q6:   Draw the refined energy profile along this path. Use the Length value for the 'reaction coordinate' axis. As the refined path for each section starts from Length = 0, add the final value of Length of the first section to the corresponding values of the second section and so on. So that you will have a continuous profile, comparable to that obtained during the initial refinement. Draw the Energy axis *relative* to the lowest energy in the system.

Q7:   Compare the final results to the initial ones. Did the refinement significantly change the energy values of the minima or the saddle-points? What was effected by the refinement? Why is the "length" of the path in the final result longer than that in the initial one?

IV. LAB WRITE-UP

Answer all the questions that were posed during the lab.

cyclo.inp

```
* Lab 10: minimum energy path for cyclohexane
*

! load topology and parameter files
OPEN READ CARD UNIT 1 NAME cycl.rtf
READ RTF   CARD UNIT 1
OPEN READ CARD UNIT 2 NAME cycl.par
READ PARAmeter CARD UNIT 2

! generate PSF
READ SEQUance CARD
* Cyclohexane
*
1
CYCL

GENErate cycl SETUp WARNing

! load the 'chair' coordinates
OPEN READ CARD UNIT 10 NAME CHAIR.CRD
READ COORdinate CARD UNIT 10
CLOSe UNIT 10

! setup non-bonded interactions
NBONd  CUTNb 150.0   CTONnb 145.0   CTOFnb 149.0   SHIFt -
       VSWItch  CDIELectric  EPSIlon 1.0
ENERgy IHBFrq 0

! print the 'chair' IC table
IC FILL
IC PRINt

! start initial optimization of reaction path
TRAVel MAXPP 100
       VERBose 0

       ! read in the two end-minima
       TRAJectory READ
              CHAIR.CRD
              TWBOAT.CRD
       DONE

       ! calculate path
       CPR NCYCle 50

       ! refine saddle point
       CPR NCYCle 100 SADDle NGRId 3 LOOPred 4

       ! get info about path
       TRAJectory ANALysis

       ! copy the coordinates of the saddle point to the main coor set
       COPY SADDle

       ! generating the full path using Steepest Descent from the
       ! saddle point
```

```
        CROSsmode
        SDP
        ! smooth and refine path
        SCM

        ! write out path as a trajectory file
        TRAJectory WRITe NAME cyclo.trj

        ! get info about path
        TRAJectory ANALysis
QUIT

! print the 'saddle-point' IC table
IC FILL
IC PRINt

vsave the saddle-point coordinates
OPEN WRITe CARD UNIT 10 NAME saddle.crd
WRITe COORdinate CARD UNIT 10

! load the 'twboat' coordinates to get IC table
OPEN READ CARD UNIT 11 NAME CHAIR.CRD
READ COORdinate CARD UNIT 11

! print the 'twboat' IC table
IC FILL
IC PRINt

END
```

initial.inp

```
* Lab 10: minimum energy paths
* Finding the energy path between two IAN conformations
* Initial path refinement
*

! load the topology and parameter files
SET t ian.rtf                    ! topology
SET p ian.par                    ! parameter
STREam toppar.str

! read the sequence from the coordinate file
OPEN UNIT 1 CARD READ NAME conf1.crd
READ SEQUence COORdinate UNIT 1
CLOSe UNIT 1

! Generate psf
GENERATE A4 SETUp

! Write psf file
OPEN UNIT 2 CARD WRITe NAME ian.psf
WRITe PSF CARD UNIT 2
* PSF file of IAN
*
CLOSe UNIT 2

! load the first conformation coordinates
OPEN READ CARD UNIT 10 NAME conf1.crd
READ COORdinate CARD UNIT 10
CLOSe UNIT 10

! setup non-bonded interactions
NBONd  CUTNb 150.0  CTONnb 145.0  CTOFnb 149.0  SHIFt-
       VSWItch  RDIElectric  EPSIlon 1.0

ENERgy IHBFrq 0

! start initial optimization of minimum energy path
TRAVel MAXPP 100
       VERBose 0
       ! read in the two conformations
       TRAJectory READ
              conf1.crd
              conf2.crd
       DONE

       ! Initial calculatation of path
       CPR NCYCle 50
       ! calculate path
       CPR NCYCle 100 SADDle
       ! get info about path
       TRAJectory ANALysis

       ! write out path as a trajectory file
       TRAJectory WRITe NAME path1.trj
QUIT
STOP
```

final.tmpl

```
* Lab 10: minimum energy paths
* Finding a conformational path between two IAN conformations
* Final refinement of a section of the complete path
*

! load the topology and parameter files
SET t ian.rtf              ! topology
SET p ian.par              ! parameter
STREam toppar.str

! load the PSF file
OPEN UNIT 1 CARD READ NAME ian.psf
READ PSF CARD UNIT 1
CLOSe UNIT 1

! load the first conformation coordinates
OPEN READ CARD UNIT 10 NAME conf1.crd
READ COORdinate CARD UNIT 10
CLOSe UNIT 10

! setup non-bonded interactions
NBONd  CUTNb 150.0  CTONnb 145.0  CTOFnb 149.0  SHIFt-
       VSWItch  RDIElectric  EPSIlon 1.0
ENERgy IHBFrq 0

! *****************************************************
! MODIFY: define the two points that define the section of
!         the path to be refined (n & m).
!         Also specify the name of the file which contains
!         the path (@trj)
! *****************************************************
SET n   ???    ! first point of the section
SET m   ???    ! last  point of the section
SET trj ???    ! name of file containing the refined path

! Second optimization of path
TRAVel  MAXPP 100
        VERBose 0

        ! read in the path from the first refinement
        TRAJectory READ NAME @trj  BEGIn @n STOP @m

        ! continue refinement
        CPR NCYCle 50 SADDle

        ! generating the full path using Steepest Descent from the
        ! saddle point
        CROSsmode
        SDP
        ! smooth and refine path
        SCM

        ! get info about path
        TRAJectory ANALysis
QUIT
STOP
```

Lab 11:   Multiple Copy Simultaneous Search

I. OBJECTIVE

In this lab you will learn about one of the methods available for computer aided drug design. Specifically, you will become familiar with the Multiple Copy Simultaneous Search (MCSS) method, analyze MCSS results, and construct peptide ligands within the HIV-1 aspartic proteinase (HIV-1 PR) binding site.

   Due to the long time it takes to run the MCSS calculations you will not perform the actual calculations but only analyze their results.

II. BACKGROUND

RATIONAL DRUG DESIGN

Rational drug design is a complex problem that is of great importance in computer-aided drug design. The goal is to predict what ligand(s) will bind most favorably to a given protein binding-site. This task involves both designing possible ligand molecules and predicting the quality of these molecule as ligands. The Multiple Copy Simultaneous Search (MCSS) method (Miranker and Karplus, 1991) is designed to tackle the first part of the job; i.e., help design ligand molecules. The idea behind MCSS is to dissect peptides into chemically "functional" groups (e.g.,

MCSS

acetamide, methanol, acetate, isobutane), search for optimal positions and orientations of these "functional" groups in the binding site, and then join together these optimal configurations to form new peptide ligands, which are optimal in the binding site. The binding constant of the candidate ligands is then estimated.

   Note, MCSS is only one of several ideas and programs that are under investigation as tools for computerized drug design.

   This lab follows the work of Caflisch, Miranker and Karplus (1993) in implementing MCSS to study the binding site of HIV-1 PR (Miller et al., 1989). In the original research the MCSS method was used to search for optimal positions and orientations of many chemically "functional" groups in this binding site and then to construct possible ligands to the protein.

N-METHYL ACETAMIDE (NMA)

In this lab you will use only the N-Methyl Acetamide groups (NMA, also called ACAM), which is a small molecule that mimics the peptide bond. You will study the NMA minima found by MCSS and see how they can be are used to construct ligand main-chains.

N-Methyl Acetamide (NMA)

NOTE: In the actual research additional types of 'functional' groups are also used (e.g., groups mimicking charged, polar or hydrophobic side-chains) and the whole data is integrated together to give a model ligand consisting of a main-chain <u>and</u> functional side-chains. Due to practical limitations this lab will focus only on main-chain prediction.

REFERENCES:

1. Miller M. et al., "Structure of complex of synthetic HIV-1 protease with a substrate-based inhibitor at 2.3 A resolution", *Science* **246** (1989), 1149-1152.
2. Miranker A. and Karplus M. "Functionality maps of binding sites: a multiple copy simultaneous search method". *Proteins* **11** (1991), 29-34.
3. Caflisch A. Miranker A. and Karplus M. "Multiple copy simultaneous search and construction of ligands in binding sites: application to inhibitors of HIV-1 aspartic proteinase", *J. Med. Chem.* **36** (1993), 2142-2167.

III. PROCEDURE

Create a new *lab11* sub-directory and change to that directory. Copy the content of the *$Lab/lab11* directory to your new directory (**cp -r** *$Lab/lab11/*** .). Type **ls** to the content of the directory.

   Among the files in this directory you will find the following three coordinate files:
i) *native.CRD* –
   The X-ray coordinates of native HIV-1 PR (resolution 2.8 Å, R-factor 0.184; Protein Data Bank entry *3hvp*)

ii) *hvp4_solv.CRD* –
The X-ray coordinates of the HIV-1 PR complex with the inhibitor MVT-101 (resolution 2.3 Å, R-factor 0.176; Protein Data Bank entry *4hvp*). These coordinates include: proteinase + inhibitor + water molecules.

iii) *hvp4.CRD* –
A strapped down version of *hvp4_solv.CRD* , which is used in the MCSS calculations. It includes only the proteinase coordinates + water molecule no. 511. Water 511 is the solvent molecule in the active site, according to the crystal structure of the complex (Miller et al., 1989).

A. STUDY THE NATIVE AND COMPLEX STRUCTURES

Use a visualization program (QUANTA, InsightII etc.) to study the structure of HIV-1 PR in the native configuration (*native.CRD*) and in the complexed configuration (*hvp4.CRD* - without the inhibitor). The highlighted commands below refer to QUANTA, but similar commands can be found in any visualization program.

Click on the **Reset View** dial button to see both structures. Since both molecules are in different orientations, you will need to superimpose them, by using the **Molecular Similarity** option in the **Applications** menu. Use the *hvp4* structure as the target (in **Manage Molecules**). and Select Residue Range 1-198 (CA only) in **Match Atoms**. Then do a **Rigid Body Fit to Target**.

Q1:   What are the main differences between native and complexed proteinase?

B. UNDERSTAND THE MCSS METHOD

Read the Miranker and Karplus (1991) paper on MCSS. Make sure you understand the method.

Q2:   Briefly point out the advantages and limitations of the MCSS method.

C. MCSS MINIMA FOR NMA

Due to practical limitations – it takes too long to run – you will not run the actual MCSS calculation.

Instead, you will find in the *lab11* directory two coordinate files that contain MCSS results for NMA (N-Methyl Acetamide) in HIV-1 PR. These results were obtained by minimizing many randomly placed copies ('replicas') of the NMA molecule in the presence of a FIXed protein.

The coordinates of the minima are sorted by increasing interaction energy, which is written in the last column:

i) *hHVP_NMA_50.CRD* – contains NMA minima obtained when starting from 5000 randomly distributed NMA groups

ii) *hHVP_NMA_1.CRD* – contains NMA minima obtained when starting from 100 randomly distributed NMA groups

Q3:   Compare the two result files (see the TIP below). How many minima were found from 100 initial groups? How many from 5000? Were all the minima of the smaller sample found in the more extensive calculation? Were the lowest minima found? Why (i.e., what interactions are stabilizing them there)?

TIP: Since every NMA group includes several atoms the above coordinate files are quite large. The comparison will be made easier if you use the UNIX **grep** command to extract only one line from each NMA group. Try typing,

grep C1 *hHVP_NMA_50.CRD* > *new_file_name*

(Remember, the second column holds the "replica number", and the last column holds the interaction energy). Do a similar **grep** on the *hHVP_NMA_1.CRD* file.

D. ANALYZE THE COMPUTED NMA MINIMA

Use a visualization program (such as QUANTA) to analyze the computed NMA minima. Load both the HIV-1 PR complexed structure (*hvp4.CRD*) and the coordinate file of the NMA minima (*hHVP_NMA_50.CRD*).

Q4: i)   Where are the NMA minima (in which proteinase subsite)?

ii) Are they clustered together? all? some?

iii) The three lowest minima are at one open end of the binding site (use color or atom selection to focus on them). Why?

Q5: i)   With which proteinase residues do the NMA minima interact?

ii) Based on the comparison you made in **Q1**, what would be the distribution of the minima if the simulations were to be carried out with the native structure of the proteinase (rather than with the complexed structure)?

Q6: i)   Give a table of the intermolecular hydrogen bonds for the 3 NMA minima characterized by the most favorable interaction energy and specify if the main-chain or a side-chain of the proteinase is involved (Hydrogen bond criterion: donor-acceptor distance < 3.5 Å and donor-H-acceptor angle > 90 degrees).

ii) Is there any correlation between the number of hydrogen bonds and the interaction energy (recall that the latter is in the last column of the file *hHVP_NMA_50.CRD*).

NOTE: The QUANTA **Calculate/Hydrogen Bond** command does not show the bonds between the hydrogen of the ACAM groups with the protein. Find these yourself by using distance and angle criteria.

E. CLUSTERING OF MCSS MINIMA

As you can see (by visual analysis with QUANTA) the NMA minimized positions tend to form clusters.

Q7:   Suggest an automatic procedure to cluster the 83 NMA minima. This procedure should take into account their interactions with the proteinase.

F. GENERATE A HEXAPEPTIDE LIGAND

Following Caflisch, Miranker and Karplus (1993), the next step is to generate sequences of possible hexapeptide ligands starting from the 83 NMA minima (one needs seven NMA minima in order to get a terminal blocked hexapeptide). We are searching for a hexapeptide ligand because the proven MVT-101 inhibitor is a hexapeptide.

By running a "connecting" program (described in section 2.2 of Caflisch et al. 1993), which connects the 83 NMA minima according to certain criteria, 8034 sequences of hexapeptides were found. One of these sequences consists of the NMA minima 15, 67, 76, 40, 20, 78 and 54. There is a corresponding CHARMM coordinates file that is named *15_67_76_40_20_78_54.CRD*. This sequence is not yet minimized and therefore it is not properly connected.

– Using what you learned in previous Labs (in particular Labs 3 and 4) write a *mini.inp* input file to minimize the connected hexapeptide (*15_67_76_40_20_78_54.CRD*). Since the MCSS calculation is performed with a fixed protein, it is wise to perform the minimization of the resulting ligands also with a FIXed protein.

Run the input file using CHARMM.

NOTE: For the purpose of this exercise the CUTOFF of the non-bonded interactions can be on the order of 5 Å.

Q8:   Explain which minimization algorithm(s) you chose, how many minimization steps were performed, and why? Record the energy before and after the minimization.

– Analyze this hexapeptide structure with QUANTA and compare the computed hexapeptide with the position of the MVT-101 inhibitor (file *hvp4_solv.CRD*).

NOTE: Although all seven elements of the hexapeptide are connected by bonds, some of these "connecting bonds" are not displayed on the QUANTA screen (because their distance is too large and the program's default setting doesn't recognize it). Nevertheless, they are still there.

Q9:   Calculate the RMS deviation between this hexapeptide and
the MVT-101 inhibitor (Use the **Molecular Similarity**
option in QUANTA; compare only the two 'ligands').

G. EVALUATING MANY HEXAPEPTIDE LIGANDS

In the *construct* sub-directory there is a special program (also named
construct) that allows you to generate a file containing the coordinates of
104 hexapeptide main chains. To execute this program move to the
construct sub-directory and type,

construct < *construct.inp* > *construct.out*

Run it.

The program will generate a coordinate file with the 104 hexapeptide
generate (*repl1.crd*). If the *construct* program does not work on your
computer architectures simply use the *repl1.crd* file that is included in the
construct sub-directory.

Analyze these sequences (file *repl1.crd*) with QUANTA. Use the
residue ID from the LAST column, and use the 'Protein with Polar
Hydrogens' topology file. Also, remove some unrealistic bonds by picking
Bond Options in the **Edit** menu, and choosing **Disable Inter-segment
Bonding**.

Q10:   Try to evaluate the orientation of the computed sequences
10 to 14, with respect to the orientation of the MVT-101
inhibitor main chain. (**TIP**: The hexapeptides are
recognized by QUANTA as "residues". Thus, sequence
number 10 is simply residue number 10 in the QUANTA
representation. Use the selection tools to focus on these
"residues").

Q11:   Analyze the hydrogen bonds of hexapeptide number 60
(sequence of NMA minima: 4, 9, 65, 13, 74, 44 and 73).

IV. WRITE-UP

Answer questions **Q1** to **Q11** above. Also answer the following question:

Q12:   The effect of solvation has been neglected during the MCSS minimization. How would you take into account the solvation effect during MCSS minimization? What would be the effect of solvation on the NMA minimized positions?

Hemoglobin Cooperativity:
the T-R Transition

I. OBJECTIVE

The objective of this lab exercise is to study the structural transition that takes place in hemoglobin upon binding of oxygen molecules (the transition from deoxy T state to the oxy R state).

The exercise is intended to give experience in some of the techniques used for detailed structural analysis in complex protein systems (~4500 atoms). Specifically you will be using the methods of (least-squares) superposition of coordinates and graphical structural examination. The *intelligent* use of atom selection, coloring and orientation in the graphics sessions will enhance the extraction of *relevant* information. The exercise is only broadly guided; it should give you a basis upon which to become familiar with hemoglobin structure.

II. BACKGROUND

DEOXY-HEMOGLOBIN (T-STATE) Hemoglobin is known to undergo a structural transition, from the deoxy T state to the oxy R state upon binding of oxygen molecules. This R-T transition stands at the core of hemoglobin cooperate activity. In this exercise we examine the structures of adult human hemoglobin (HbA) in the deoxy (T) and oxy (R) states. By comparison of the structures we observe the changes that occur on oxygenation of hemoglobin, both in tertiary structure (within a subunit) and in quaternary structure (between subunits). This will allow us to visualize components of the proposed structural mechanism of hemoglobin cooperativity.

OXY-HEMOGLOBIN (R-STATE)

It is assumed that you are familiar with basic aspects of the overall construction of hemoglobin (8 helices per subunit, the heme, the terms *proximal* and *distal*, α and β subunits and subunit assembly). It will also be convenient if you are familiar with the standard globin residue naming convention. For details of the above see for example R.E. Dickerson's book *Hemoglobin* (Benjamin Cummings, 1983) or an article by Fermi and Perutz on Hemoglobin and Myoglobin in *Atlas of Biological Structures* (Phillips and Richards Eds, Clarendon, 1981) or many other texts....

The unveiling of Hemoglobin's cooperative mechanism has come from the study of a number of different hemoglobins, crystallized under different conditions and with various ligands bound to them. Due to their different ligand saturation levels and quaternary structural arrangement,

213

some of these hemoglobins represent intermediate "states" between the *classic* R and *classic* T structures. Here, for simplicity, we restrict ourselves to two structures: a deoxy R conformation and a fully oxygenated T conformation. These are sufficient to illustrate the main structural phenomena.

REFERENCES:

For discussions of detailed aspects of the structural differences and the proposed cooperative mechanism see:

1.   Perutz M.E.,"Stereochemistry of cooperative effects in haemoglobin", *Nature* **228** (1970), 726-734.
2.   Gelin B.R. and Karplus M., "Mechanism of tertiary structural change in hemoglobin", *PNAS (USA)* **74** (1977), 801-805.
3.   Baldwin J. and Chothia C., "Haemoglobin: the structural changes related to ligand binding and its allosteric mechanism", *J. Mol. Biol.* **129** (1979), 175-220.
4.   Gelin B.R., Lee A.W., and Karplus M.,"Hemoglobin tertiary structural change on ligand binding: its role in the co-operative mechanism", *J. Mol. Biol.* **171** (1983), 489-559.
5.   Perutz M.F., Fermi G., Luisi B., Shaanan B., and Liddington R.C, "Stereochemistry of cooperative mechanisms in hemoglobin", *Acc. Chem. Res.* **20** (1987), 309-317.

NOTE: It is strongly advised that you read some of these references **before** tackling the lab.

This lab exercise is divided into four sections, in which you will
(a)   Become acquainted with the overall arrangement of the molecule (the α-1 subunit, the α-1/β-1 dimer and the tetramer).
(b)   Learn how to superimpose structures in appropriate reference frame for comparison.
(c)   Study tertiary structure changes (in the α-1 and β-1 subunits).
(d)   Study quaternary structure changes(in the α-1/β-2 interface and in the α-1/α-2 and β-1/β-2 interfaces).

III. PROCEDURE

Create a new sub-directory *lab12* and change to that directory. Copy the content of the *$Lab/lab12* directory to your own directory (**cp** *$Lab/lab12/* .*). Move to the new directory (**cd** *lab12*) and view its content (**ls**). You should see two CHARMM coordinate files: *oxy.CRD* and *deoxy.CRD*.

The X-ray coordinates sets that you will use in this lab are:

Deoxy (T): Deoxy HbA at 1.74 Å resolution (from the PDB entry *2hhb*). Reference: Fermi G. et al. *J. Mol. Biol.* **175** (1984), 159.

Oxy (R):   HbA with oxygen bound at 2.1 Å resolution. (from the PDB entry *1hho*). Reference: Shaanan, B. *J. Mol. Biol.* **171** (1983), 31.

These coordinate files are in the CHARMM format, each containing a tetramer. The files have the following features:
– Only protein heavy-atoms (no hydrogens, no solvent and no counterions).

– The segment identifiers are:
  alp1 and *bet1* for α-1 and β-1 subunits respectively
  alp2 and *bet2* for α-2 and β-2 subunits respectively

– Each subunit has the residue identifiers according to the globin name convention: e.g., residue His 58 in α-1 is named RESId E7 and residue His 92 in β-1 is named RESId F8 (in QUANTA's selection convention they are reached by: "*zone alp1:e7*" and "*zone bet1:f8*" respectively). These naming schemes should make navigation and comparison easier.

You will be using a visualization program, such as QUANTA, throughout this exercise. The following highlighted commands are in QUANTA notation, but similar commands exist is all visualization programs. If you are using a program other than QUANTA see that you make the proper adaptations.

Begin by opening a session in your visualization program (e.g., QUANTA).

TIP: It may be of use to create a QUANTA view file (under **View**) to save orientations that prove to be useful. This can be very helpful in navigating through the structures.

A. OVERALL ARRANGEMENT OF THE MOLECULE

1) The α-1 subunit:

**α-1
SUBUNIT**

– Load (**Import**) either the *oxy* or *deoxy* structure (RESID from LAST column).
– Create an atom selection for the α-1 subunit (*seg alp1*). You can use the **Type in a Selection** option (although there are other ways to carry out this selection).
– Create a display selection to show the backbone (N, Ca, C) atoms of the globin, the heme and the proximal histidine (*alp1:f8*) of the α-1 subunit. One possible way is to type in a user defined selection such as,

> *pick c ca n # seg alp1*
> *zone alp1:f8 alp1:heme*

– Choose a suitable color scheme to distinguish globin from heme!
– Examine the structure. Look at the arrangement of the helices, the position of the heme and the heme-globin linkage.
– Change the display selection to view all of the α-1 subunit atoms (too much information?).
– Change the display selection to view all residues with atoms within 7Å of the ND1 atom of the distal histidine (E7) (e.g., type in: *round 7. ND1 alp1:E7*). Color backbone/sidechain distinctly.

Q1:   List the residues that comprise the distal environment (and their type). Note, when you click on a residue the type will appear in the Textport.

2) The α-1/β-1 dimer:

**α-1/β-1
SUBUNIT**

– Modify the selections to include both the α-1 and β-1 subunit backbone atoms (color accordingly).
– Use **Manage View/Orient** from the **View** menu to look along the +X axis (The axis are shown in the upper left corner).

Q2:   Which secondary structure elements of the subunits make up the interface?

3) The tetramer:

TETRAMER

- Modify the above selections to examine the complete tetramer.
- Use an appropriate color selection to distinguish α and βsubunits.
- Look at the arrangement of the subunits and the various subunit interfaces (try orthogonal views, -Z for the α-1/β-2 interface).

Q3:   What are the distances between the iron atoms in the 4 subunits? (Get a feeling for the dimensions of this molecular assembly).

B. SUPERPOSITION OF STRUCTURES IN THE BGH FRAME

BGH VIEW

As the structural differences of R and T involve both tertiary and quaternary structure changes it is essential that a suitable reference frame be selected for their comparison (why?). Baldwin and Chothia showed that the structure in the region of contact at the α-1/β-1 interface is the same in R and T structures. If R and T structures are superposed in this region, they showed that the differences between α-1 and β-1 reflect tertiary structural changes and that differences between α-2 and β-2 reflect both tertiary and quaternary structural changes. The region for superposition includes residues from B, G and H helices and thus this particular frame of reference is known as BGH, and is commonly used.

- Load both oxy and deoxy structures (RESID from LAST column).
- Create an atom selection for all atoms in both structures, change display to show only backbone.
- Make each structure a different color.
- Under the **Applications** menu, choose **Molecular Similarity** (it may take some time to load and open its new pallet).
- Make deoxy the **Target Molecule** by clicking on its name in the *Molecule Management* window and than selecting **Define Target Molecule** from the *Molecular Similarity* pallet. The asterisk that appears next to the molecule's name indicates it has been selected (also see the *Textport*).
- Transform the molecules into the GBH frame by superposing the α-carbons of the following residues: B4-B14, G6-G18 and H5-H17, in both α-1 (*alp1*) and β-1 (*bet1*) subunits. Use **Match Atoms** and then **Select Residue Ranges** (use **C Alpha Only** option, and remember to **Append** the various matches). In the interactive box that appears type the residue range (e.g., *alp1:b4*) both in the "Fixed Frame" slot and in the "Moving Frame" slots. You will have to perform 6 residue range specifications, giving a total of 74 atom matches. You can see the lines

connecting the matched atoms if you choose **View/Stamp**. Use **View/Mono** to return to normal view.
- Reward your typing by saving the atom matches to a file, and then **Exit Match Atoms**.
- **Selecting Rigid Body Fit to Target** performs the superposition.

Q4:   What is the RMS coordinate difference in this region (see *Textport*)?

- Write out the transformed oxy coordinates. These will be used in subsequent comparisons.

TIP: The transformation that you have applied remains current until a new atom selection is made. Thus you can stay within **Molecular Similarity** and just change display selections. If you move out, then either load your saved transformed coordinates or re-apply your transformation.

C. TERTIARY STRUCTURE CHANGES

Examining the T/R differences in the individual subunits. Use the transformed oxy coordinates (or stay within the comparison application).

1) The α-1 subunit:

**α-1
SUBUNIT**
- Create a display selection for the α-1 subunits in both the deoxy and oxy states, to show the backbone atoms of the globin and the heme.
- Flash between the two structures describe the differences between the hemes. Either use the **Flash Molecule** option from the *Molecular Similarity* palette or manually flash between them from within the *Molecule Management* window (the **Visible** option).

Q5:   What is the behavior of the iron atom? Which regions of backbone differ most?

- Change the display to the following: Show the backbone of the globin from residue E1 to residue G1, the Heme and show in full details residues F8, FG3, FG5, C7, CD1, CD4, E7 and E11 (if you want include the ligand in the oxy structure (type *O2*)). The displayed backbone serves as a reference frame for the specific residues we like to focus on.

A useful selection here is:

> *pick n ca c # zone alp1:e1 to alp1:g1*
> *zone alp1:f8 alp1:fg3 alp1:fg5*
> *zone alp1:c7 alp1:cd1 alp1:cd4 alp1:e7 alp1:e11*
> *zone alp1:heme alp1:o2*

- Notice the movement of the proximal histidine (F8) associated with the heme.
- Notice the movement in the sidechains of Leu FG3 and Val FG5.
- Notice how the structure of the distal residues is largely the same in the two structures.

Q6:   The heme, His F8, the H-helix and the FG corner have been described as the *allosteric core* of hemoglobin (Gelin, Lee and Karplus, 1983). Why ?

2) The β-1 subunit:

β-1
SUBUNIT

- Arrange for the selection of the residues specified above showing the allosteric core and distal residues in the β-1 subunit.

Q7:   What are the major differences you observe from the α-1 subunit?

D. QUATERNARY STRUCTURE CHANGES

Examining the T/R differences between the subunits. Again use the transformed oxy coordinates (or stay within the comparison application).

1) The α-1/β-2 interface:

α-1/β-2
INTERFACE

- Create a display selection to show atoms of α-1 and β-2 subunits.
- Color the FG corners, C helices and HC termini of each subunit the same, but distinct from one another. Make everything else the same color.
- Use **Manage View/Orient** from the **View** menu to look along the z-axis.

– See how the FG corner of α-1 and the C helix of β-2 remain similar in packing arrangement and how the C helix of α-1 and the FG corner of β-2 differ. These regions have been described as the *joint* and *switch* of the quaternary structure changes.

Q8:   To what specific structural feature does the term *switch* refer? Can you see it? How do the quaternary shifts effect specific interactions at the interface, how is this important in the cooperative mechanism?

α-1/α-2   2) The α-1/α-2 and β-1/β-2 interfaces
β-1/β-2
INTERFACE – Put up the whole tetramer.

Q9:   What is the major effect upon going from T to R on the central cavity of the tetramer?

TIP: Flash between the two structures.

IV. WRITE-UP

Answer all the questions, **Q1** through **Q9** presented throughout the lab. In addition answer:

Q10:   What is an allosteric effector? Give examples and describe their principal binding sites.

REFERENCES

General references:

1. Brooks B. R., Bruccoleri R. E., Olafson B. D., States D. J., Swaminathan S., Karplus M., "CHARMM: A Program for Macromolecular Energy, Minimization, and Dynamics Calculations". *J. Comput. Chem.* (1983), **4**:187-217.

2. Brooks C.L. III, Karplus M., Pettitt B.M. *Proteins: A Theoretical Perspective of Dynamics, Structure and Thermodynamics* (John Wiley & Sons: New York, 1988).

3. McCammon J.A., Harvey S.C. *Dynamics of Proteins and Nucleic Acids* (Cambridge University Press: Cambridge, 1987).

4. Becker O.M., MacKerell A.D. Jr., Roux B., Watanabe M. (Eds.), *Computational Biochemistry and Biophysics* (Marcel Dekker: New York, 2001).

5. Allen M.P., Tildesley D. J., *Computer Simulations of Liquids*, 2nd edition (Oxford University Press: Oxford, 1989).

CHARMM References:

1. Brooks B.R., Bruccoleri R.E., Olafson B.D., States D.J., Swaminathan S., Karplus M. "CHARMM: A Program for Macromolecular Energy, Minimization, and Dynamics Calculations," *J. Comp. Chem.* (1983) **4**, 187-217.

2. MacKerell A.D. Jr., Brooks B., Brooks C.L. III, Nilsson L., Roux B., Won Y., Karplus M. "CHARMM: The Energy Function and Its Parameterization with an Overview of the Program," in: *The Encyclopedia of Computational Chemistry*, **1**, 271-277, P. v. R. Schleyer *et al.*, editors (John Wiley & Sons: Chichester, 1998).

Lab Specific References:

Lab 2

1. Lavie A., Allen K.N , Petsko G.A., Ringe D. "X-ray crystallographic structures of D-xylose isomerase-substrate complexes position the substrate and provide evidence for metal movement during catalysis." *Biochemistry* (1994) **33**:5469-5480.

Lab 4

2. Cotrait M., Kreissler M., Hoflack J., Lehn J.-M., Maigret B. "Computational simulations of the conformational behavior of the adhesive proteins RGDS fragment." *J. Computer-Aided Mol. Des.* **6** (1992), 113-130.

Lab 5

3. Brooks C.L. III, Karplus M., "Solvent effects on protein motion and protein effects on solvent motion. Dynamics of the active site region of lysozyme." *J. Mol. Biol.* (1989), **208**:159-181.
4. Pomès R., McCammon J.A., "Mass and step length optimization for the calculation of equilibrium properties by molecular dynamics simulation " *Chem. Phys. Lett.* (1990), **166**: 425-428.

Lab 6

5. van Gunsteren W.F., Berendsen H.J.C., "Algorithms for macromolecular dynamics and constraint dynamics." *Mol. Phys.* (1977) **34**: 1311-1327.
6. van Gunsteren W. F., Karplus M., "Effect of constraints on the dynamics of macromolecules." *Macromolecules* (1982) **15**: 1528-1544.

Lab 7

7. Case D.A., Karplus M. "Dynamics of ligand binding to heme proteins." *J. Mol. Biol.* (1979), **132**: 343-368.
8. Kottalam J., Case D.A., "Dynamics of ligand escape from the heme pocket of myoglobin." *J. Am. Chem. Soc.* (1988), **110**: 7690-7697.

9. Elber R., Karplus M., "Enhanced sampling in molecular dynamics: use of the time-dependent Hartree approximation for a simulation of carbon monoxide diffusion through myoglobin." *J. Am. Chem. Soc.* (1990), **112**: 9161-9175.

10. Takano, T. "Structure of myoglobin refined at 2.0 A resolution. I. Crystallographic refinement of metmyoglobin from sperm whale." *J Mol Biol.* (1977) **110**:537-568.

Lab 8

11. Goldstein H., *Classical Mechanics* (Addison-Welsley 1980), Ch. 6.

12. Levitt M., Sander C., Stern P.S., " Protein normal-mode dynamics: Trypsin inhibitor, crambin, ribonuclease and lysozyme ." *J. Mol. Biol.* (1985) **181**:423-447.

13. Brooks B., Karplus M., "Harmonic Dynamics of Proteins: Normal Modes and Fluctuations in Bovine Pancreatic Trypsin Inhibitor." *Proc. Natl. Acad. Sci. USA* (1983) **80**:6571-6575.

14. Go N., Noguti T., Nishikawa T., "Dynamics of a Small Globular Protein in Terms of Low-Frequency Vibrational Modes." *Proc. Natl. Acad. Sci. USA* (1983) **80**:3696-3700.

15. Brooks B., Karplus M., "Normal Modes for Specific Motions of Macromolecules: Application to the Hinge-Bending Mode of Lysozyme." *Proc. Natl. Acad. Sci. USA* (1985) **82**:4995-4999.

16. Simonson T., Perahia D., "Normal modes of symmetric protein assemblies. Application to the tobacco mosaic virus protein disk." *Biophys. J.* (1992) **61**:410-427.

Lab 9

17. Kollman P., "Free Energy Calculations: Applications to Chemical and Biochemical Phenomena", *Chem. Rev.* (1993) **93**:2395-2417.

18. Zwanzig R.W., "High-Temperature Equation of State by a Perturbation Method. I. Nonpolar Gases", *J. Chem. Phys.* (1954) **22**:1420-1426.

19. Kirkwood J.G., "Statistical Mechanics of fluid Mixtures", *J. Chem. Phys.* (1935) **3**:300-313.

20. Boresch S., Archontis G., Karplus M., "Free energy simulations - The meaning of the individual Contributions from a Component Analysis", *Proteins* (1994) **20**:25-33.

21. Brady G. P., Sharp K. A., "Decomposition of interaction free energies in proteins and other complex systems." *J Mol Biol.* (1995) **254**:77-85.

Lab10

22. Fischer S., Karplus M., "Conjugate peak refinement: an algorithm for finding reaction paths and accurate transition states in systems with many degrees of freedom." *Chem. Phys. Lett.* (1992) **194**:252-261.
23. Elber R., Karplus M., "A method for determining reaction paths in large molecules: Application to myoglobin." *Chem. Phys. Lett.* (1987) **139**:375-380.
24. Ulitsky A., Elber R., "A new technique to calculate steepest descent paths in flexible polyatomic systems." *J. Chem. Phys.* (1990), **92**:1510-1511.

Lab11

25. Miller M., Schneider J., Sathyanarayana B.K., Toth M.V., Marshall G.R., Clawson L., Selk L., Kent S.B., Wlodawer A. "Structure of complex of synthetic HIV-1 protease with a substrate-based inhibitor at 2.3 A resolution." *Science* (1989) **246**:1149-1152.
26. Miranker A., Karplus M., "Functionality maps of binding sites: a multiple copy simultaneous search method." *Proteins* (1991) **11**:29-34.
27. Caflisch A., Miranker A., Karplus M., "Multiple copy simultaneous search and construction of ligands in binding sites: application to inhibitors of HIV-1 aspartic proteinase." *J. Med. Chem.* **36** (1993), 2142-2167.

Lab12

28. Dickerson R.E., Geiss I. *Hemoglobin: structure, function, evolution, and pathology* (Benjamin Cummings, Menlo Park, 1983).
29. Fermi G., Perutz M.F. "Hemoglobin and Myoglobin", in: *Atlas of Biological Structures*. D.C. Phillips and F.M. Richards Eds, (Clarendon, 1981).
30. Perutz M.F., "Stereochemistry of cooperative effects in haemoglobin." *Nature* (1970) **228**:726-739.
31. Gelin B.R., Karplus M., "Mechanism of Tertiary Structural Change in Hemoglobin." *Proc. Natl. Acad. Sci. USA* (1977) **74**:801-805.
32. Baldwin J., Chothia C., "Haemoglobin: The structural changes related to ligand binding and its allosteric mechanism." *J. Mol. Biol.* (1979) **129**:175-220.

33. Gelin B.R., Lee A.W-M., Karplus M., "Hemoglobin tertiary structural change on ligand binding its role in the co-operative mechanism." *J. Mol. Biol.* (1983) **171**:489-559.

34. Perutz M.F., Fermi G., Luisi B., Shaanan B., Liddington R.C. "Stereochemistry of cooperative mechanisms in hemoglobin." *Accs. Chem. Res.* (1987) **20**:309-317.

35. Fermi G., Perutz M.F., Shaanan B., Fourme R. "The crystal structure of human deoxyhaemoglobin at 1.74 Å resolution." *J. Mol. Biol.* (1984) **175**:159-174.

36. Shaanan B. "Structure of human oxyhaemoglobin at 2.1 A resolution." *J. Mol. Biol.* (1983) **171**:31-59.

PDB FILES

Crystallography derived protein structures are used in various labs throughout the book. Following are their PDB entries and primary references (file names in the book are specified when different from the PDB entry code). Additional information on each of these x-ray structures can be found on the PDB web site (http://www.rcsb.org/pdb/).

3FAB
Saul F.A., Poljak R.J. "Crystal structure of human immunoglobulin fragment Fab New refined at 2.0 A resolution." *Proteins* (1992) **14**:363-371.

2HHB (file: deoxy.crd)
Fermi G., Perutz M.F., Shaanan B., Fourme R. "The crystal structure of human deoxyhaemoglobin at 1.74 Å resolution." *J. Mol. Biol.* (1984) **175**:159-174.

1HHO (file: oxy.crd)
Shaanan, B. "Structure of human oxyhaemoglobin at 2.1 A resolution." *J. Mol. Biol.* (1983) **171**:31-59.

3HVP (file: native.CRD)
Wlodawer A., Miller M., Jaskolski M., Sathyanarayana B.K., Baldwin E., Weber I.T., Selk L.M., Clawson L., Schneider J., Kent S.B.

"Conserved folding in retroviral proteases: crystal structure of a synthetic HIV-1 protease." *Science* (1989) **245**:616-621.

4HVP
Miller M., Schneider J., Sathyanarayana B.K., Toth M.V., Marshall G.R., Clawson L., Selk L., Kent S.B., Wlodawer A. "Structure of complex of synthetic HIV-1 protease with a substrate-based inhibitor at 2.3 A resolution." *Science* (1989) **246**:1149-1152.

8LYZ
Beddell C.R., Blake C.C., Oatley S.J. "An x-ray study of the structure and binding properties of iodine-inactivated lysozyme." *J.Mol.Biol.* (1975) **97**: 643-654.

9LYZ
Kelly J.A., Sielecki A.R., Sykes B.D., James M.N., Phillips D.C. "X-ray crystallography of the binding of the bacterial cell wall trisaccharide NAM-NAG-NAM to lysozyme." *Nature* (1979) **282**:875-878.

1MBC (file: mbco_pdb4.crd)
Kuriyan J., Wilz S., Karplus M., Petsko G.A. "X-ray structure and refinement of carbon-monoxy (Fe II)-myoglobin at 1.5 A resolution." *J.Mol.Biol.* (1986) **192**:133-154.

2MBN (file: dxmbx.crd)
Takano T. "Structure of myoglobin refined at 2.0 A resolution. I. Crystallographic refinement of metmyoglobin from sperm whale." *J Mol Biol.* (1977) **110**:537-568.

1MBO
Phillips S.E. "Structure and refinement of oxymyoglobin at 1.6 A resolution." *J.Mol.Biol.* (1980) **142**:531-554.

1XYA (file: wtxi.pdb)
Lavie A., Allen K.N , Petsko G.A., Ringe D. "X-ray crystallographic structures of D-xylose isomerase-substrate complexes position the substrate and provide evidence for metal movement during catalysis." *Biochemistry* (1994) **33**:5469-5480.

HOW TO OBTAIN CHARMM (CHARMm) AND QUANTA for
USERS OF THIS GUIDE

For users of the Guide in academia, a six-month free license for
CHARMm and QUANTA can be obtained by contacting Accelrys at
<drisal@accelrys.com>.

This access is available either for user of the Guide enrolled in a course,
or for studying on their own as part of an academic institution.

Users of the Guide in commercial settings will have to make individual
arrangements for access to CHARMm and QUANTA.

A list of local Accelrys contacts is available at
http://www.accelrys.com/about /find.html

1,4-diamino-butane (DAB) 177-178
5-amino-valeric acid (AVA) 177-178

Adopted-Basis Newton-Raphson 56
Amino-butane (AB) 177-178
Autocorrelation function 103-104

β-turns 66
Benzene 159-164
BGH view 217
Boundary conditions 79, 90
    Periodic Boundary conditions 79

Canonical partition function 173
CHARMM 10-17
    Command line 13
    Running jobs 14-15
    Variables 13
CHARMM commands
    ABNR 56
    CDIElectric 54
    Comments 14
    CONJ 56
    COOR 107, 110, 135, 139
    COORdinate 71
    COPY 107
    CORRelation 112
    CRD 12
    DEFIne 106
    DYNAmics 82
    ENERgy 72
    GENErate 71
    GRMS 57
    HBUIld 106, 132
    IC 12, 72, 111
    INTEraction 107
    MINImize 73
    NBONd 71
    PARAmeters 11, 71
    PATCh 105
    PED 160-161

    PERTubation 179
    PSF 11
    RDIElectric 55
    RTF 11, 71
    SCALar 110
    SD 55, 73
    SELEct 105
    SET 13
    SHAKe 109
    SHIFt 53
    STREam 14
    SWITch 52
    TRAJectory 139
    TRAVel 195
    VIBRan 160
Conformational analysis 65-73
Conjugated Gradients 56
Conjugated Peak Refinement (CPR) 194
Constraints 39-40
Cyclohexane 41-45, 195-197

Dihedral angles 36

Free energy calculations 173-185
    Backwards integration 182
    Canonical partition function 173
    Effect of random seed 182
    Exponential Formula 174-175
    Helmholtz free energy 173
    Hysteresis 177
    Individual contributions 184-185
    Path dependence 177, 183-184
    Thermodynamic perturbation 174, 175-176
Fourier Transform 104

Glucose 45-49
    α-glucose 45-46
    D-glucose 46-49

Harmonic 40
Helmholtz free energy 173

Hemoglobin 213-220
  α1 subunit 215, 218
  α1/β1 dimer 216, 219
  β1 subunit 219
  BGH view 217
  Deoxy-hemoglobin (T state) 213
  Oxy-hemoglobin (R state) 213
  Tetramer 216
HIV-1 protease 207-208
Hysteresis 177

IC Table 103
IgG Fab 32
Improper dihedral angles 36
Iisobutyryl-(ala)3-NH-methyl (IAN)
  197-198

Ligand dynamics 134-141
Lysozyme 58-59

Maxwell-Boltzmann distribution 76
MCSS 205-211
MeOH 90-93
Minimization 38-39, 55-57
  Adopted-Basis Newton-Raphson 56
  Conjugated Gradients 56
  Gradient RMS 56
  Newton-Raphson 56
  Steepest descent 55
Minimum energy path 193-194
Molecular dynamics 75-80
  Boundary conditions 79, 90
  Convergence 79
  dynamics trajectory 67-68
  Integrators 76-78
  Newton's equation of motion 75
  Procedure 76
  Solvent effect 79
  Temperature 76
  Verlet 77-78
Molecular models 19-20
  All atom model 20
  Extended atom model 20
  Polar hydrogen model 20
Molecular visualization 23-30

MSF format 21
Multiple Copy Simultaneous Search
  (MCSS) 205-211
Myoglobin 32, 101, 105, 131, 133

Newton-Raphson 56
Newton's equation of motion 75
N-methyl acetamide (NMA) 205-206
Normal Mode Analysis 155-164
  Harmonic approximation 156
  Normal modes 156
  Vibrational Spectrum 156
  Degeneracy 156, 161-162
  IR activity 157, 162
  Limitation 157
  Analysis 161
  In-plane/out-of-plane 162, 163

Parameters 37-38
Periodic boundary condition 79
Phenol 165
Potential energy function 35-38
  Angle 35
  Bond 35
  Bonded interactions 35-36
  Constant 54
  Cutoff 52
  Dielectric model 54
  Distance dependant 55
  Electrostatic 37
  Non-bonded interactions 37
  SHIFT 53
  SWITCH 52
  Torsion 36
  Truncation schemes 51-52
  vdW 37

QUANTA 21-32

Rational drug design 205-211
Reaction path 193
RGD peptide 65, 80-81
RMS 57
R-T transition in hemoglobin 213-220

SHAKE 40, 101-102
SHIFT 53
Solvent 79
Steepest descent 55
SWITCH 52

TIP3P water model 79, 90
Transition state 194

UNIX 4-9

Vacuum simulation 80
Verlet 77-78

Xylose Isomerase 46

FOCUS ON STRUCTURAL BIOLOGY

1. S.R. Kiihne and H.J.M. de Groot (eds.): *Perspectives on Solid State NMR in Biology.* 2001
    ISBN 0-7923-7102-X
2. J.N. Housby (ed.): *Mass Spectrometry and Genomic Analysis.* 2001   ISBN 0-7923-7173-9
3. C. Dennison: *A Guide to Protein Isolation.* 2003   ISBN 1-4020-1224-1
4. O.M. Becker and M. Karplus (eds.): *Guide to Biomolecular Simulations.* 2005
    ISBN 1-4020-3586-1